崧燁文化

曹永忠、許智誠、蔡英德　著

Arduino程式教學(溫溼度模組篇)

Arduino Programming (Temperature& Humidity Modules)

自序

Arduino 系列的書出版至今，已經過三年，出書量也近八十本大關，當初出版電子書是希望能夠在教育界開一門 Maker 自造者相關的課程，沒想到一寫就已過三年，繁簡體加起來的出版數也已也近八十本的量，這些書都是我學習當一個 Maker 累積下來的成果。

這本書可以說是我的書另一個里程碑，很久以前，這個系列開始以駭客的觀點為主，希望 Maker 可以擁有駭客的觀點、技術、能力，駭入每一個產品設計思維，並且成功的重製、開發、超越原有的產品設計，這才是一位對社會有貢獻的『駭客』。

如許多學習程式設計的學子，為了最新的科技潮流，使用著最新的科技工具與軟體元件，當他們面對許多原有的軟體元件沒有支持的需求或軟體架構下沒有直接直持的開發工具，此時就產生了莫大的開發瓶頸，這些都是為了追求最新的科技技術而忘卻了學習原有基礎科技訓練所致。

筆著鑒於這樣的困境，思考著『如何駭入眾人現有知識寶庫轉換為我的知識』的思維，如果我們可以駭入產品結構與設計思維，那麼了解產品的機構運作原理與方法就不是一件難事了。更進一步我們可以將原有產品改造、升級、創新，並可以將學習到的技術運用其他技術或新技術領域，透過這樣學習思維與方法，可以更快速的掌握研發與製造的核心技術，相信這樣的學習方式，會比起在已建構好的開發模組或學習套件中學習某個新技術或原理，來的更踏實的多。

目前許多學子在學習程式設計之時，恐怕最不能了解的問題是，我為何要寫九九乘法表、為何要寫遞迴程式，為何要寫成函式型式…等等疑問，只因為在學校的學子，學習程式是為了可以了解『撰寫程式』的邏輯，並訓練且建立如何運用程式邏輯的能力，解譯現實中面對的問題。然而現實中的問題往往太過於複雜，授課的老師無法有多餘的時間與資源去解釋現實中複雜問題，期望能將現實中複雜問題淬鍊成邏輯上的思路，加以訓練學生其解題思路，但是眾多學子宥於現實問題的困惑，無法單純用純粹的解題思路來進行學習與訓練，反而以現實中的複雜來反駁老師教學太過學理，沒有實務上的應用為由，拒絕深入學習，這樣的情形，反而自己造成了學習上的障礙。

本系列的書籍，針對目前學習上的盲點，希望讀者當一位產品駭客，將現有產品的產品透過逆向工程的手法，進而了解核心控制系統之軟硬體，再透過簡單易學的 Arduino 單晶片與 C 語言，重新開發出原有產品，進而改進、加強、創新其原有產品固有思維與架構。如此一來，因為學子們進行『重新開發產品』過程之中，可以很有把握的了解自己正在進行什麼，對於學習過程之中，透過實務需求導引著開發過程，可以讓學子們讓實務產出與邏輯化思考產生關連，如此可以一掃過去陰霾，更踏實的進行學習。

這三年多以來的經驗分享，逐漸在這群學子身上看到發芽，開始成長，覺得 Maker 的教育方式，極有可能在未來成為教育的主流，相信我每日、每月、每年不斷的努力之下，未來 Maker 的教育、推廣、普及、成熟將指日可待。

最後，請大家可以加入 Maker 的 Open Knowledge 的行列。

曹永忠 於貓咪樂園

自序

記得自己在大學資訊工程系修習電子電路實驗的時候，自己對於設計與製作電路板是一點興趣也沒有，然後又沒有天分，所以那是苦不堪言的一堂課，還好當年有我同組的好同學，努力的照顧我，命令我做這做那，我不會的他就自己做，如此讓我解決了資訊工程學系課程中，我最不擅長的課。

當時資訊工程學系對於設計電子電路課程，大多數都是專攻軟體的學生去修習時，系上的用意應該是要大家軟硬兼修，尤其是在台灣這個大部分是硬體為主的產業環境，但是對於一個軟體設計，但是缺乏硬體專業訓練，或是對於眾多機械機構與機電整合原理不太有概念的人，在理解現代的許多機電整合設計時，學習上都會有很多的困擾與障礙，因為專精於軟體設計的人，不一定能很容易就懂機電控制設計與機電整合。懂得機電控制的人，也不一定知道軟體該如何運作，不同的機電控制或是軟體開發常常都會有不同的解決方法。

除非您很有各方面的天賦，或是在學校巧遇名師教導，否則通常不太容易能在機電控制與機電整合這方面自我學習，進而成為專業人員。

而自從有了 Arduino 這個平台後，上述的困擾就大部分迎刃而解了，因為 Arduino 這個平台讓你可以以不變應萬變，用一致性的平台，來做很多機電控制、機電整合學習，進而將軟體開發整合到機構設計之中，在這個機械、電子、電機、資訊、工程等整合領域，不失為一個很大的福音，尤其在創意掛帥的年代，能夠自己創新想法，從 Original Idea 到產品開發與整合能夠自己獨立完整設計出來，自己就能夠更容易完全了解與掌握核心技術與產業技術，整個開發過程必定可以提供思維上與實務上更多的收穫。

Arduino 平台引進台灣自今，雖然越來越多的書籍出版，但是從設計、開發、製作出一個完整產品並解析產品設計思維，這樣產品開發的書籍仍然鮮見，尤其是能夠從頭到尾，利用範例與理論解釋並重，完完整整的解說如何用 Arduino 設計出一個完整產品，介紹開發過程中，機電控制與軟體整合相關技術與範例，如此的書籍更是付之闕如。永忠、英德兄與敝人計畫撰寫 Maker 系列，就是基於這樣對市場需要的觀察，開發出這樣的書籍。

作者出版了許多的 Arduino 系列的書籍，深深覺的，基礎乃是最根本的實力，所以回到最基礎的地方，希望透過最基本的程式設計教學，來提供眾多的 Makers 在入門 Arduino 時，如何開始，如何攥寫自己的程式，進而介紹不同的週邊模組，主要的目的是希望學子可以學到如何使用這些週邊模組來設計程式，期望在未來產品開發時，可以更得心應手的使用這些週邊模組與感測器，更快將自己的想法實現，希望讀者可以了解與學習到作者寫書的初衷。

許智誠　　於中壢雙連坡中央大學 管理學院

自序

隨著資通技術(ICT)的進步與普及，取得資料不僅方便快速，傳播資訊的管道也多樣化與便利。然而，在網路搜尋到的資料卻越來越巨量，如何將在眾多的資料之中篩選出正確的資訊，進而萃取出您要的知識？如何獲得同時具廣度與深度的知識？如何一次就獲得最正確的知識？相信這些都是大家共同思考的問題。

為了解決這些困惱大家的問題，永忠、智誠兄與敝人計畫製作一系列「Maker 系列」書籍來傳遞兼具廣度與深度的軟體開發知識，希望讀者能利用這些書籍迅速掌握正確知識。首先規劃「以一個 Maker 的觀點，找尋所有可用資源並整合相關技術，透過創意與逆向工程的技法進行設計與開發」的系列書籍，運用現有的產品或零件，透過駭入產品的逆向工程的手法，拆解後並重製其控制核心，並使用 Arduino 相關技術進行產品設計與開發等過程，讓電子、機械、電機、控制、軟體、工程進行跨領域的整合。

近年來 Arduino 異軍突起，在許多大學，甚至高中職、國中，甚至許多出社會的工程達人，都以 Arduino 為單晶片控制裝置，整合許多感測器、馬達、動力機構、手機、平板...等，開發出許多具創意的互動產品與數位藝術。由於 Arduino 的簡單、易用、價格合理、資源眾多，許多大專院校及社團都推出相關課程與研習機會來學習與推廣。

以往介紹 ICT 技術的書籍大部份以理論開始、為了深化開發與專業技術，往往忘記這些產品產品開發背後所需要的背景、動機、需求、環境因素等，讓讀者在學習之間，不容易了解當初開發這些產品的原始創意與想法，基於這樣的原因，一般人學起來特別感到吃力與迷惘。

本書為了讀者能夠深入了解產品開發的背景，本系列整合 Maker 自造者的觀念與創意發想，深入產品技術核心，進而開發產品，只要讀者跟著本書一步一步研習與實作，在完成之際，回頭思考，就很容易了解開發產品的整體思維。透過這樣的思路，讀者就可以輕易地轉移學習經驗至其他相關的產品實作上。

所以本書是能夠自修的書，讀完後不僅能依據書本的實作說明準備材料來製作，盡情享受 DIY(Do It Yourself)的樂趣，還能了解其原理並推展至其他應用。有興趣的讀者可再利用書後的參考文獻繼續研讀相關資料。

本書的發行有新的創舉，就是以電子書型式發行，在國家圖書館(http://www.ncl.edu.tw/)、國立公共資訊圖書館 National Library of Public Information(http://www.nlpi.edu.tw/)、台灣雲端圖庫(http://www.ebookservice.tw/)等都可以免費借閱與閱讀，如要購買的讀者也可以到許多電子書網路商城、Google Books 與 Google Play 都可以購買之後下載與閱讀。希望讀者能珍惜機會閱讀及學習，繼續將知識與資訊傳播出去，讓有興趣的眾人都受益。希望這個拋磚引玉的舉動能讓更多人響應與跟進，一起共襄盛舉。

本書可能還有不盡完美之處，非常歡迎您的指教與建議。近期還將推出其他 Arduino 相關應用與實作的書籍，敬請期待。

最後，請您立刻行動翻書閱讀。

蔡英德 於台中沙鹿靜宜大學主顧樓

目 錄

Maker 系列

本書是『Arduino 程式教學』的第九本書，主要是給讀者熟悉 Arduino 的溫度、濕度周邊模組的介紹、使用方式、電路連接範例等等。

Arduino 開發板最強大的不只是它的簡單易學的開發工具，最強大的是它封富的周邊模組與簡單易學的模組函式庫，幾乎 Maker 想到的東西，都有廠商或 Maker 開發它的周邊模組，透過這些周邊模組，Maker 可以輕易的將想要完成的東西用堆積木的方式快速建立，而且最強大的是這些周邊模組都有對應的函式庫，讓 Maker 不需要具有深厚的電子、電機與電路能力，就可以輕易駕御這些模組。

所以本書要介紹市面上最常見、最受歡迎與使用的顯示模組，讓讀者可以輕鬆學會這些常用模組的使用方法，進而提升各位 Maker 的實力。

CHAPTER

熱敏電阻

本章主要介紹常用於偵測溫度的熱敏電阻，主要是讓讀者瞭解，溫度是如何被量出來的。

顧名思義，熱敏電阻的電阻值，隨著溫度的變化而改變，與一般的固定電阻不同。屬於可變電阻的一類。熱敏電阻是開發早、種類多、發展較成熟的敏感元器件。熱敏電阻由半導體陶瓷材料組成，熱敏電阻利用的原理是溫度引起電阻變化。

熱敏電阻

熱敏電阻英文全名 Thermally Sensitive Resistance，簡稱 TSR。顧名思義，它是一種對溫度（熱)相當敏感的電阻，有時亦稱為熱阻體（Thermistor)(參考下圖所示)。

TSR 乃以半導體氧化物燒結而成，是一種能被大量生的產品，在一般度要求不高的溫度量測或溫度控制等場一般而言，是一種非常方便測量溫度的感測器。

首先，對於溫度測量方面，若讀者不熟悉，可以參閱拙作『Arduino 電風扇設計與製作』(曹永忠, 许智诚, & 蔡英德, 2014; 曹永忠, 許智誠, & 蔡英德, 2013)、『Arduino 飲水機電子控制器開發: The Development of a Controller for Drinking Fountain Using Arduino』(曹永忠, 許智誠, & 蔡英德, 2014a, 2014b)，有興趣讀者可到 Google Books (https://play.google.com/store/books/author?id=曹永忠) & Google Play (https://play.google.com/store/books/author?id= 曹 永 忠) 或 Pubu 電子書城 (http://www.pubu.com.tw/store/ultima) 購買該書閱讀之。

 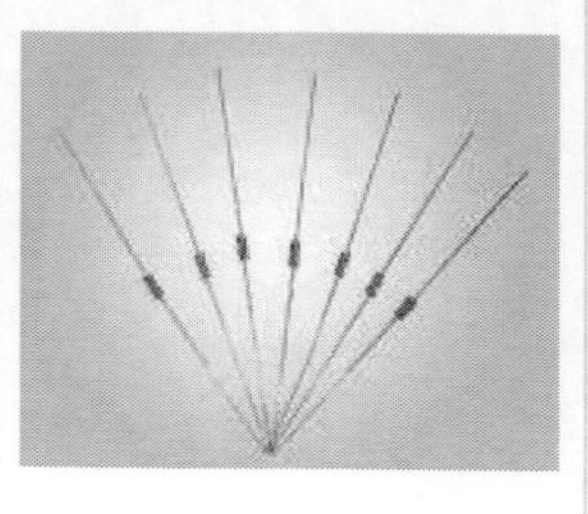 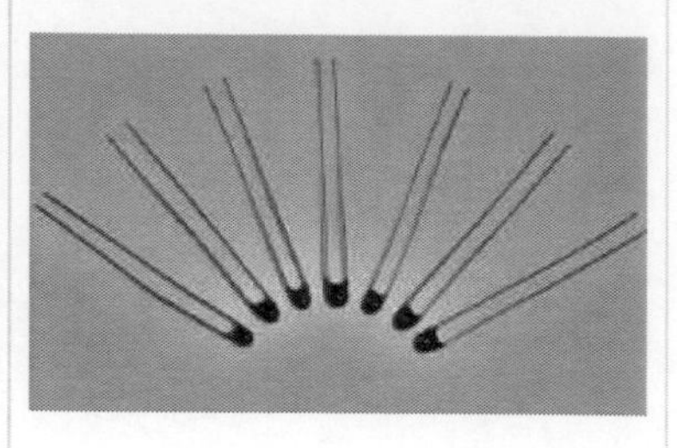

圖 1 常見熱敏電阻

首先我們先來介紹熱敏電阻(TSR)的特性：

TSR 因材質的不同，大至可分成三類，分別為正溫度係數（Positive Temperature Coefficient ：PTC)、負溫度係數（Negative Temperature Coefficient ：NTC)及臨界溫度係數（Critical Temperature Coefficient ：CTC)等，可以參考圖 2 所示)，下列為每一種熱敏電阻(TSR)的特性的大概簡介。

- PTC：正溫度係數（Positive Temperature Coefficient)，此種 TSR 會隨溫度增加，使電阻增大。
- NTC：負溫度係數（Negative Temperature Coefficient)，此種 NTC 會隨溫度增加，使電阻減少。
- CTC：臨界溫度係數（Critical Temperature Coefficient)，只針對某一特定的溫度範圍內，該類之電阻會迅速的變化。

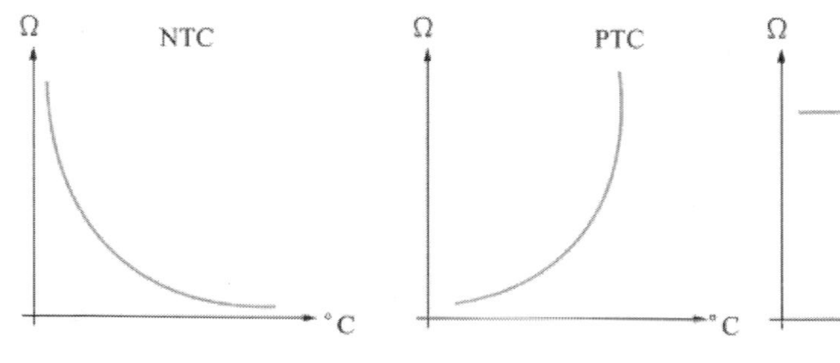

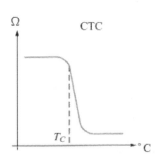

圖 2 常見熱敏電阻特性圖

熱敏電阻模組

許多地方我們都需要量測溫度，所以使用溫度感測模組是最普通不過的事，我們本節介紹溫度感測模組（如圖 9 所示），它主要是使用溫度感應電組作成溫度感測模組。

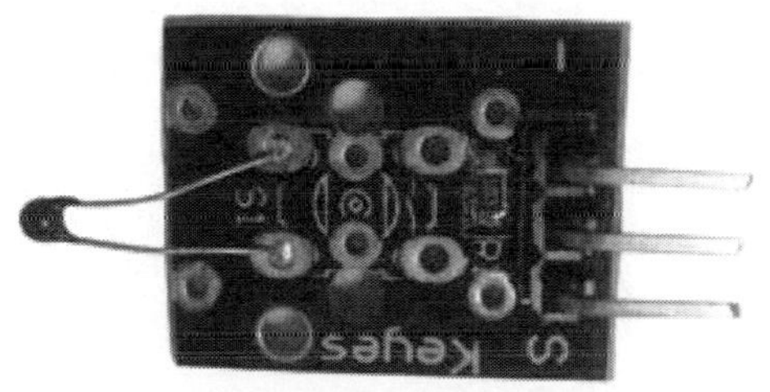

圖 3 熱敏電阻模組

本實驗是採用熱敏電阻模組，如上圖所示，先參考下圖所示之熱敏電阻模組腳位圖腳，在遵照下表熱敏電阻模組接腳表進行電路組裝。

圖 4 類熱敏電阻模組腳位圖

表 1 熱敏電阻模組接腳表

接腳	接腳說明	Arduino 開發板接腳
S	Vcc	電源（+5V) Arduino +5V
2	GND	Arduino GND

接腳	接腳說明	Arduino 開發板接腳
3	Signal	Arduino analog pin A1

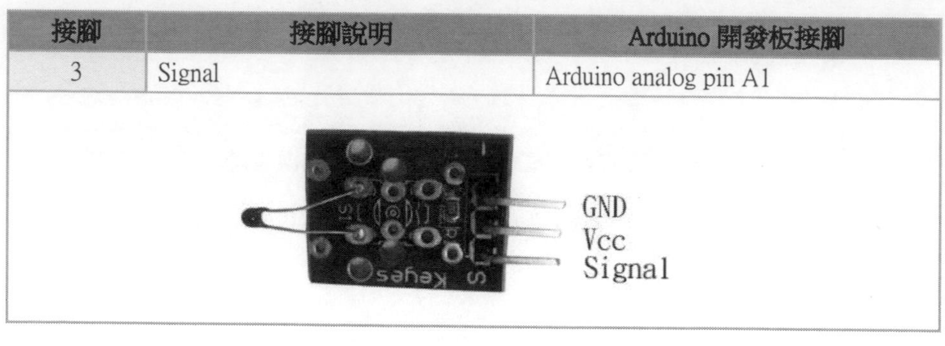

我們遵照前幾章所述，將 Arduino 開發板的驅動程式安裝好之後，我們打開 Arduino 開發板的開發工具：Sketch IDE 整合開發軟體(軟體下載請到：https://www.arduino.cc/en/Main/Software)，編寫一段程式，如下表所示之類比溫度傳感器模組測試程式。

表 2 類比溫度傳感器模組測試程式

類比溫度傳感器模組測試程式(Temp_sensor)

```
#include <LiquidCrystal.h>
#define DPin 7
#define LedPin 6
#define APin A0

 LiquidCrystal lcd(8, 9, 10, 45, 43, 41,39,37,35,33,31);

    int val = 0 ;
    int val1 = 0 ;
 void setup()
{
pinMode(LedPin,OUTPUT);//設置數位 IO 腳模式，OUTPUT 為 Output
 pinMode(DPin,INPUT);//定義 digital 為輸入介面
 //pinMode(APin,INPUT);//定義為類比輸入介面



   Serial.begin(9600);//設定串列傳輸速率為 9600 }

 // set up the LCD's number of columns and rows:
```

```
  lcd.begin(16, 2);
  // Print a message to the LCD.
  lcd.print("Vibration Sensor ");
}
void loop() {

  // set the cursor to column 0, line 1
  // (note: line 1 is the second row, since counting begins with 0):
   val=analogRead(APin);//讀取感測器的值
   val1=digitalRead(DPin);//讀取感測器的值
    Serial.print(val);//輸出模擬值，並將其列印出來
    Serial.print("/");//輸出模擬值，並將其列印出來
    Serial.print(val1);//輸出模擬值，並將其列印出來
    Serial.print("\n");//輸出模擬值，並將其列印出來

 delay(100);
}
```

程式下載：https://github.com/brucetsao/eTemperature_Humidity

當然，如圖 5所示，我們可以看到類比溫度傳感器模組結果畫面。

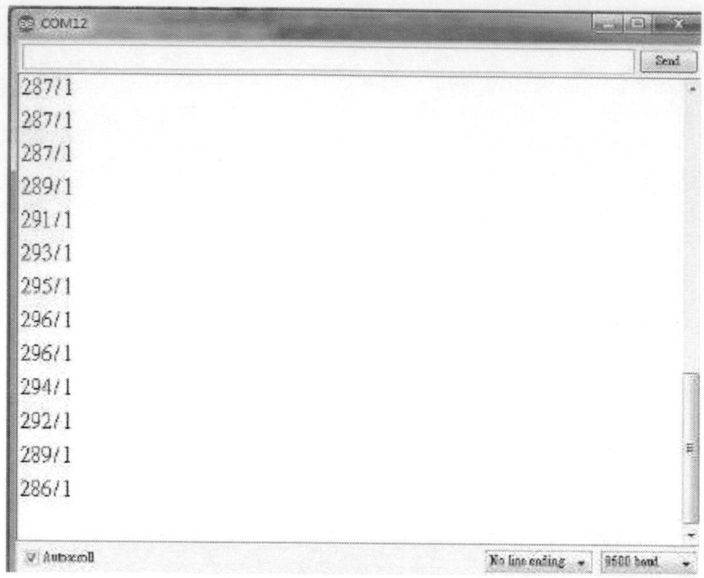

圖 5 類比溫度傳感器模組結果畫面

章節小結

本章主要介紹之 Arduino 開發板使用與連接 OLED 顯示模組，透過本章節的解說，相信讀者會對連接、使用 OLED 顯示模組，有更深入的了解與體認。

2
CHAPTER

LM35 溫度感測器

在溫度的量測領域中，感測溫度的方式有多種型態，依特性可概分為膨脹變化型、顏色變化型、電阻變化型、電流變化型、電壓變化型、頻率變化型…等，常見的電壓變化型的溫度感測器有 LM35 溫度感測器、其不同點為 LM35 溫度感測器之輸出電壓是與攝氏溫標呈線性關係，而 LM335 溫度感測器則是與凱氏溫標呈線性關係。由於攝氏溫標較常使用，因此本章節將針對 LM35 溫度感測器為本章主題。

溫度感測模組(LM35)

LM35 溫度感測器是很常用且易用的溫度感測器元件，在元器件的應用上也只需要一個 LM35 元件，只利用一個類比介面就可以，將讀取的類比值轉換為實際的溫度，其接腳的定義，請參考下圖.(c) LM35 溫度感測器所示。

所需的元器件如下。

- 直插 LM35*1
- 麵包板*1
- 麵包板跳線*1 紮

如下圖所示，這個實驗我們需要用到的實驗硬體有下圖.(a)的 Arduino Mega 2560 與下圖.(b) USB 下載線、下圖.(c) LM35 溫度感測器、下圖.(d).LCD1602 液晶顯示器：

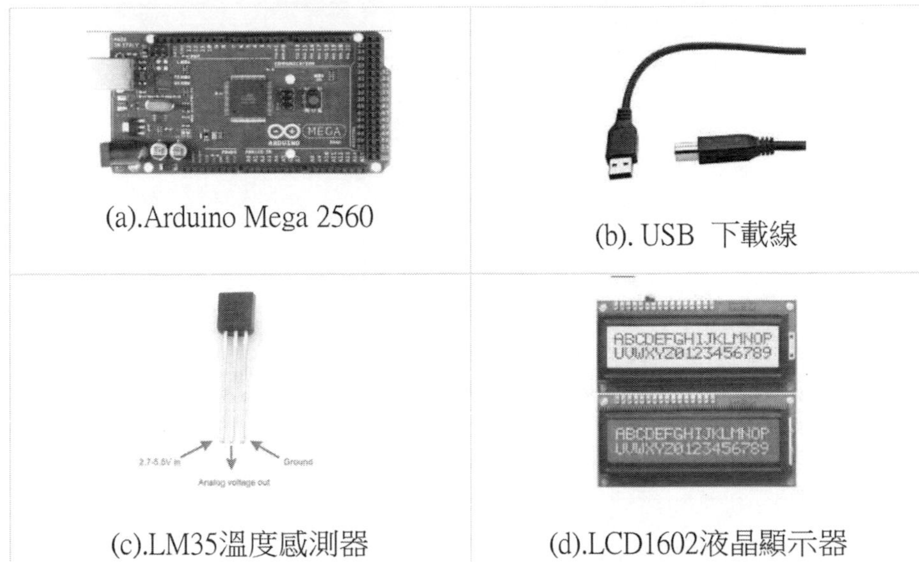

(a).Arduino Mega 2560

(b). USB 下載線

(c).LM35溫度感測器

(d).LCD1602液晶顯示器

圖 6 LM35 溫度感測器所需材料表

讀者可以參考下圖所示之 LM35 溫度感測連接電路圖，進行電路組立。

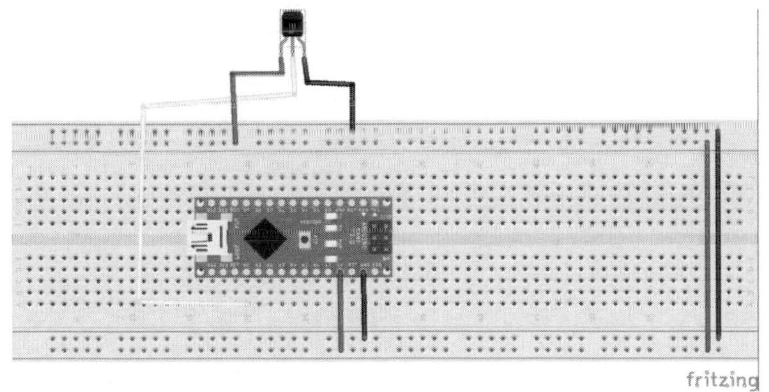

圖 7 LM35 溫度感測連接電路圖

讀者也可以參考下表之腳位說明，進行電路組立。

表 3 溫度感測模組(LM35)接腳表

接腳	接腳說明	Arduino 開發板接腳

接腳	接腳說明	Arduino 開發板接腳
S	Vcc	電源 (+5V) Arduino +5V
2	GND	Arduino GND
3	Signal	Arduino analog Pin 0

接腳	接腳說明	接腳名稱
1	Ground (0V)	接地 (0V) Arduino GND
2	Supply voltage; 5V (4.7V – 5.3V)	電源 (+5V) Arduino +5V
3	SDA	Arduino SDA Pin
4	SCL	Arduino SCL Pin21

資料來源：Arduino 程式教學(入門篇):Arduino Programming (Basic Skills & Tricks)

(曹永忠, 許智誠, & 蔡英德, 2015a)

我們遵照前幾章所述，將 Arduino 開發板的驅動程式安裝好之後，我們打開 Arduino 開發板的開發工具：Sketch IDE 整合開發軟體(軟體下載請到：https://www.arduino.cc/en/Main/Software)，攥寫一段程式，如下表所示之 LM35 溫度感測器程式程式，讓 Arduino 讀取 LM35 溫度感測器程式，並把溫度顯示在監控畫面與 LCD1602 液晶顯示器上。

表 4 LM35 溫度 IC 感測器程式

LM35 溫度 IC 感測器程式(LM35)

```
// include the library code:
#include <Wire.h>
#include <LiquidCrystal_I2C.h>
// initialize the library with the numbers of the interface Pins
```

```
LiquidCrystal_I2C lcd(0x27, 16, 2); // set the LCD address to 0x27 for a 16 chars and 2 line display

int potPin = 0; //定義類比介面 0 連接 LM35 溫度感測器
void setup()
{
Serial.begin(9600);//設置串列傳輸速率
 // set up the LCD's number of columns and rows:
  lcd.begin(16, 2);
  // Print a message to the LCD.

}
void loop()
{
int val;//定義變數
int dat;//定義變數
val=analogRead(0);// 讀取感測器的模擬值並賦值給 val
dat=(125*val)>>8;//溫度計算公式
Serial.print("Tep:");//原樣輸出顯示 Tep 字串代表溫度
Serial.print(dat);//輸出顯示 dat 的值
Serial.println("C");//原樣輸出顯示 C 字串
 // set the cursor to column 0, line 1
  // (note: line 1 is the second row, since counting begins with 0):
  lcd.setCursor(0, 1);
      lcd.print("Tep:");
      lcd.print(dat);
      lcd.print(" .C");
delay(500);//延時 0.5 秒
}
```

程式下載：https://github.com/brucetsao/eTemperature_Humidity

讀者也可以在作者 YouTube 頻道 (https://www.youtube.com/user/UltimaBruce) 中，在網址 https://www.youtube.com/watch?v=rTk5gCBfYI4&feature=youtu.be，看到本次實驗-

LM35 溫度感測器程式結果畫面。

當然、如下圖所示，我們可以看到 LM35 溫度感測器程式結果畫面。

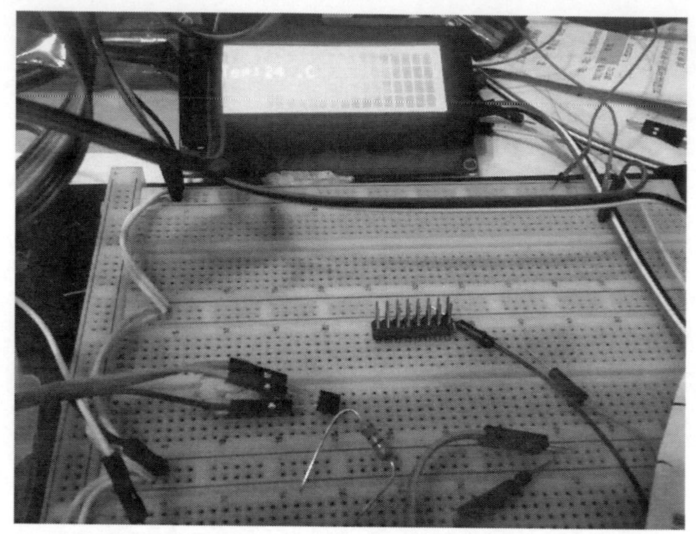

圖 8 LM35 溫度感測器程式結果畫面

章節小結

本章主要介紹之 Arduino 開發板使用與連接 LM35 溫度感測器，透過本章節的解說，相信讀者會對連接、使用 LM35 溫度感測器來量測溫度，有更深入的了解與體認。

CHAPTER

DS18B20 數位溫度感測器

許多地方我們都需要量測溫度，所以使用溫度感測模組是最普通不過的事，我們本節介紹溫度感測模組(DS18B20) (如下圖所示)，它主要是使用 DS18B20 溫度感測器作成溫度感測模組(DS18B20)。

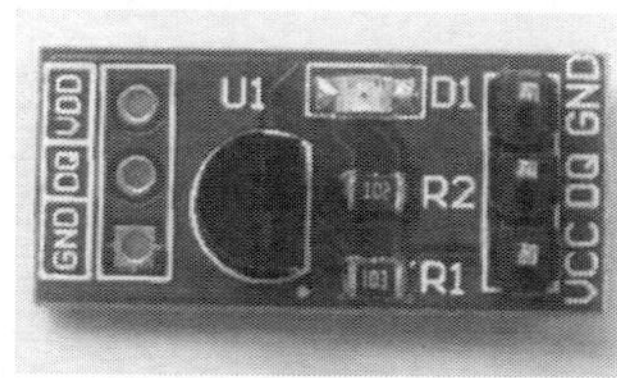

圖 9 溫度感測模組(DS18B20)

DS18B20 溫度感測模組提供 高達 9 位元溫度準確度來顯示物品的溫度。而溫度的資料只需將訊號經過單線串列送入 DS18B20 或從 DS18B20 送出，因此從中央處理器到 DS18B20 僅需連接一條線（和地）(如下下圖所示)。

DS18B20 溫度感測模組讀、寫和完成溫度變換所需的電源可以由數據線本身提供，而不需要外部電源。因為每一個 DS18B20 溫度感測模組有唯一的系列號（silicon serial number），因此多個 DS18B20 溫度感測模組可以存在於同一條單線總線上。這允許在許多不同的地方放置 DS18B20 溫度感測模組。

(a). DS-18B20

PIN ASSIGNMENT

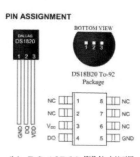

BOTTOM VIEW

DS18B20 To-92 Package

NC 1 8 NC
NC 2 7 NC
V_{DD} 3 6 NC
DQ 4 5 GND

GND DQ VDD

(b). DS-18B20 腳位說明

(c). DS-18B20 防水型

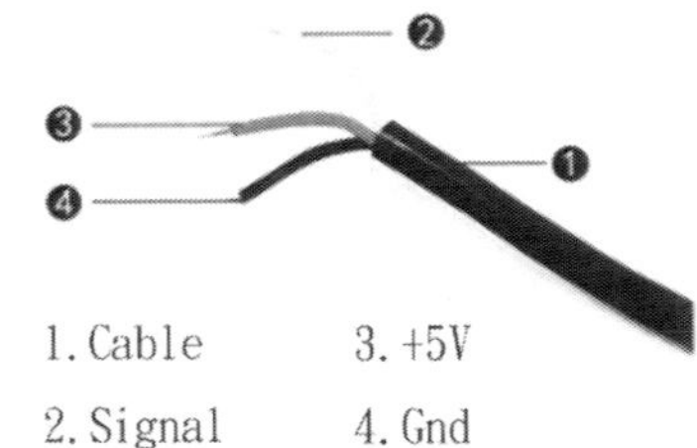

(d). DS-18B20 防水型腳位說明

圖 10 DS-18B20 數位溫度感測器

DS-18B20 數位溫度感測器特性介紹

1. DS18B20 的主要特性

- 適應電壓範圍更寬，電壓範圍：3.0～5.5V，在寄生電源方式下可由數據線供電
- 獨特的單線介面方式，DS18B20 在與微處理器連接時僅需要一條口線即可實現微處理器與 DS18B20 的

2. 雙向通訊

- DS18B20 支援多點組網功能，多個 DS18B20 可以並聯在唯一的三線上，實現組網多點測溫
- DS18B20 在使用中不需要任何週邊元件，全部傳感元件及轉換電路集成在形如一只三極管的積體電路內
- 可測量溫度範圍為－55℃～＋125℃，在-10～+85℃時精度為±0.5℃
- 程式讀取的解析度為 9～12 位元，對應的可分辨溫度分別為 0.5℃、0.25℃、0.125℃和 0.0625℃，可達到高精度測溫
- 在 9 位元解析度狀態時，最快在 93.75ms 內就可以把溫度轉換為數

位資料，在 12 位元解析度狀態時，最快在 750ms 內把溫度值轉換為數位資料，速度更快

- 測量結果直接輸出數位溫度信號，只需要使用一條線路的資料匯流排，使用串列方式傳送給微處理機，並同時可傳送 CRC 檢驗碼，且具有極強的抗干擾除錯能力
- 負壓特性：電源正負極性接反時，晶片不會因發熱而燒毀， 只是不能正常工作。

3. DS18B20 的外形和內部結構

- DS18B20 內部結構主要由四部分組成：64 位元 ROM 、溫度感測器、非揮發的溫度報警觸發器 TH 和 T 配置暫存器。
- DS18B20 的外形及管腳排列如圖 11 所示

4. DS18B20 接腳定義：(如下圖所示)。

- DQ 為數位資號輸入/輸出端；
- GND 為電源地；
- VDD 為外接供電電源輸入端。

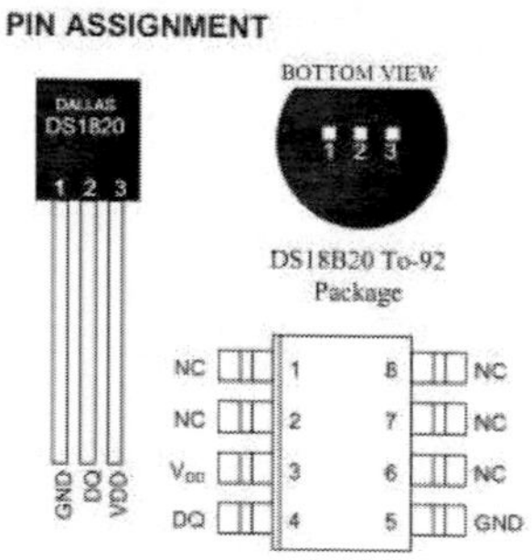

圖 11 DS18B20 腳位一覽圖

本實驗是採用溫度感測模組(DS18B20)，如上圖所示，先參考下圖所示之溫度感測模組(DS18B20)腳位圖腳，在遵照下表所示之溫度感測模組(DS18B20 接腳表進行電路組裝。

圖 12 溫度感測模組(DS18B20)腳位圖

讀者可以參考下圖所示之 DS18B20 溫度感測連接電路圖，也可以參考下表之腳位說明，進行電路組立。

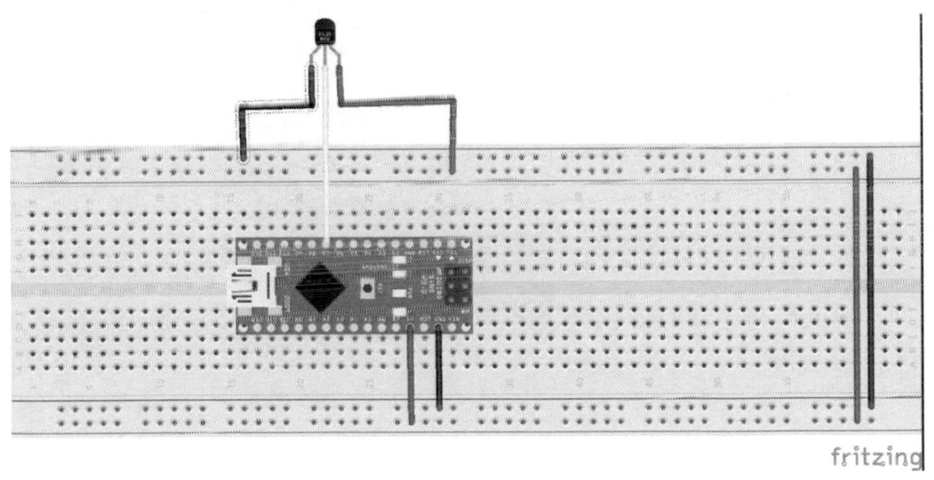

圖 13　DS18B20 溫度感測連接電路圖

表 5 溫度感測模組(DS18B20)接腳表

接腳	接腳說明	Arduino 開發板接腳
S	Vcc	電源 (+5V) Arduino +5V
2	GND	Arduino GND
3	Signal	Arduino digital Pin 7

接腳	接腳說明	Arduino 開發板接腳

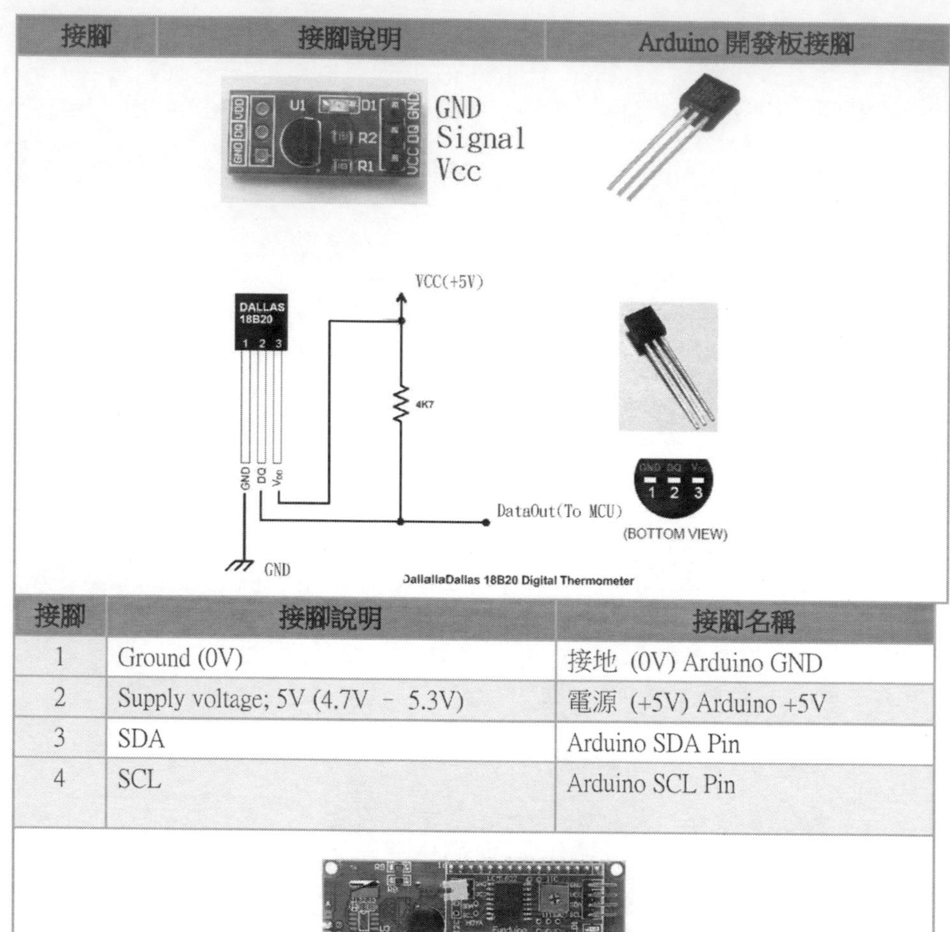

接腳	接腳說明	接腳名稱
1	Ground (0V)	接地 (0V) Arduino GND
2	Supply voltage; 5V (4.7V – 5.3V)	電源 (+5V) Arduino +5V
3	SDA	Arduino SDA Pin
4	SCL	Arduino SCL Pin

資料來源：Arduino 程式教學(入門篇):Arduino Programming (Basic Skills & Tricks) (曹永忠, 許智誠, et al., 2015a; 曹永忠, 許智誠, & 蔡英德, 2015b, 2015c, 2015d)

我們遵照前幾章所述，將 Arduino 開發板的驅動程式安裝好之後，我們打開 Arduino 開發板的開發工具：Sketch IDE 整合開發軟體(軟體下載請到：https://www.arduino.cc/en/Main/Software)，攥寫一段程式，如下表所示之溫度感測模組(DS18B20)測試程式。

表 6 溫度感測模組(DS18B20)測試程式

溫度感測模組(DS18B20)測試程式(DS18B20)

```
#include <Wire.h>
#include <LiquidCrystal_I2C.h>
#include <OneWire.h>
#include <DallasTemperature.h>
#define ONE_WIRE_BUS 7

LiquidCrystal_I2C lcd(0x27, 16, 2); // set the LCD address to 0x27 for a 16 chars and 2
line display
OneWire oneWire(ONE_WIRE_BUS);
DallasTemperature sensors(&oneWire);

void setup(void)
{
  Serial.begin(9600);
  Serial.println("Temperature Sensor");
    lcd.begin(16, 2);
  // Print a message to the LCD.
  lcd.print("DallasTemperature");

  // 初始化
  sensors.begin();
}

void loop(void)
{
  // 要求匯流排上的所有感測器進行溫度轉換
  sensors.requestTemperatures();

  // 取得溫度讀數（攝氏）並輸出，
  // 參數 0 代表匯流排上第 0 個 1-Wire 裝置
  Serial.println(sensors.getTempCByIndex(0));
  lcd.setCursor(1, 1);
    lcd.print("                    ") ;
   lcd.setCursor(1, 1);
  lcd.print(sensors.getTempCByIndex(0));
```

```
  delay(2000);
}
```

讀者也可以在作者 YouTube 頻道 (https://www.youtube.com/user/UltimaBruce)中，在網址 https://www.youtube.com/watch?v=HqcWcVTkHKA&feature=youtu.be，看到本次實驗-溫度感測模組(DS18B20)測試程式結果畫面。

當然、如下圖所示，我們可以看到溫度感測模組(DS18B20)測試程式結果畫面。

圖 14 溫度感測模組(DS18B20)測試程式結果畫面

DallasTemperature 函式庫介紹

Arduino 開發版驅動 DS18B20 溫度感測模組，需要 DallasTemperature 函數庫，而 DallasTemperature 函數庫則需要 OneWire 函數庫，讀者可以在本書附錄中找到這些函市庫，也可以到作者 Github(https://github.com/brucetsao)網站中，在筆者 Arduino 專用的函示庫目錄 https://github.com/brucetsao/LIB_for_MCU/tree/master/Arduino_Lib/libraries 下載到

DallasTemperature、OneWire 等函數庫。

下列簡單介紹 DallasTemperature 函式庫內各個函式的解釋與用法：

- uint8_t getDeviceCount(void)，回傳 1-Wire 匯流排上有多少個裝置。
- typedef uint8_t DeviceAddress[8]，裝置的位址。
- bool getAddress(uint8_t*, const uint8_t)，回傳某個裝置的位址。
- uint8_t getResolution(uint8_t*)，取得某裝置的溫度解析度（9~12 bits，分別對應 0.5℃、0.25℃、0.125℃、0.0625℃），參數為位址。
- bool setResolution(uint8_t*, uint8_t)，設定某裝置的溫度解析度。
- bool requestTemperaturesByAddress(uint8_t*)，命令某感測器進行溫度轉換，參數為位址。
- bool requestTemperaturesByIndex(uint8_t)，同上，參數為索引值。
- float getTempC(uint8_t*)，取得溫度讀數，參數為位址。
- float getTempCByIndex(uint8_t)，取得溫度讀數，參數為索引值。
- 另有兩個靜態成員函式可作攝氏華氏轉換。
    - static float toFahrenheit(const float)
    - static float toCelsius(const float)

章節小結

本章節內容主要是解釋如何使用 DS18B20 溫度感測模組的功能，所以我們必須了解目前常用的溫度感測器使用方法，方能繼續往下實作，繼續進行我們的實驗。

CHAPTER

白金感溫電阻(PT100)

一般金屬對於電流皆具有阻力，其阻值亦皆會隨溫度的變化而改變，改變的情形將因材質而異，電阻式溫度感測器(Resistance Temperature Detector：RTD)即是利用此一特性以金屬細線繞製而成，所以電阻式溫度感測器(Resistance Temperature Detector：RTD)均具有正溫度係數，其中以白金、銅、鎳等材質最普遍。

白金細線所繞製而成的感溫電組具有最高的精確度及安定性，尤其在-200℃至600℃之間其電組-溫度的線性變化，遠比銅、鎳等材質好很多，目前用的最多的為Pt 100 白金感溫電阻。

白金感溫電阻

白金感溫電阻(PT100)具有高精確度及高安定性，在-200℃~600℃之間亦有很好的線性度。一般而言，白金感溫電阻(PT100)感溫電阻在低溫-200℃~-100℃間其溫度係數較大；在中溫 100℃~300℃間有相當良好的線性特性；而在高溫 300℃~500℃間其溫度係數則變小。由於在 0℃時，白金感溫電阻(PT100)電阻值為 100Ω，已被視為金屬感溫電阻的標準規格。

白金感溫電阻(PT100)感溫電阻使用時應避免工作電流太大，以減低自體發熱，因此可限制其額定電流在 2 mA 以下。由於白金感溫電阻(PT100)自體發熱 1mW 約會造成 0.02℃~0.75℃的溫度變化量，所以降低白金感溫電阻(PT100)的電流亦可降低其溫度變化量。然而，若電流太小，則易受雜訊干擾，所以一般白金感溫電阻(PT100)之電流以限制在 0.5mA~2mA 間為宜。

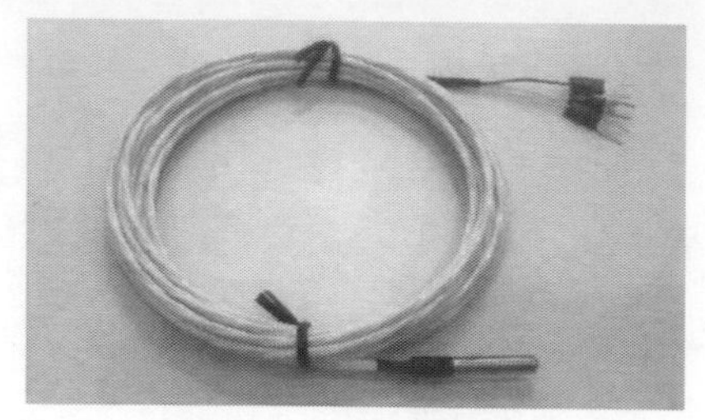

圖 15 三線式白金感溫電阻(PT100)一覽圖

白金感溫電阻(PT100)感溫電阻值與溫度間之關係式，可表亦為：

(2)低溫-200℃~0℃間：

方程式 1 白金感溫電阻(PT100) 低溫電阻值與溫度間之關係式

$$R(T) = R(0) \bullet [1 + 3.90802 \times 10^{-3} \bullet T - 0.580195 \times 10^{-6} \bullet T^2 - 4.27350 \times 10^{-12} \bullet (T - 100)T^3]$$

(2)高溫 0℃~500℃間：

方程式 2 白金感溫電阻(PT100) 高溫電阻值與溫度間之關係式

$$R(T) = R(0) \bullet [1 + 3.90802 \times 10^{-3} \bullet T - 0.580195 \times 10^{-6} \bullet T^2$$

而對於白金 Pt100 感溫電阻與溫度間之關係式，由於其在 0℃時之電阻值為

方程式 3 白金感溫電阻(PT100) 0℃時之電阻值

$$R(0) = 10 \times 10^2 = 1K$$

故白金感溫電阻(PT100)電阻值與溫度間之關係式為 ：

方程式 4 白金感溫電阻(PT100)電阻值與溫度間之關係式

$$R(T) \approx 1\text{k}\Omega + 3.90802\ \Omega / ^\circ\text{C} \cdot \text{T}\ ^\circ\text{C}$$

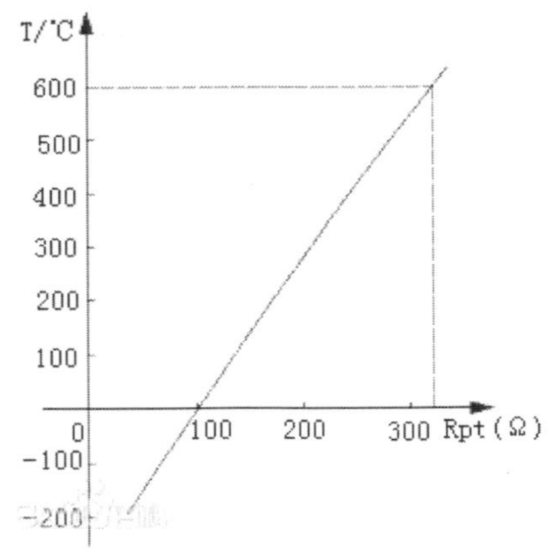

圖 16 PT100 溫度阻值表(重畫)

資料來源：百度百科：http://baike.baidu.com/view/1299879.htm

MAX6675 K 型熱電偶感測器

為了可以使用 K 型熱電偶感測器，我們使用如下圖所示之 Max6675 整合型的晶片來處理 K 型熱電偶感測器，我們依照如下表所示，將 MAX6675 K 型熱電偶感測器模組電路組立。

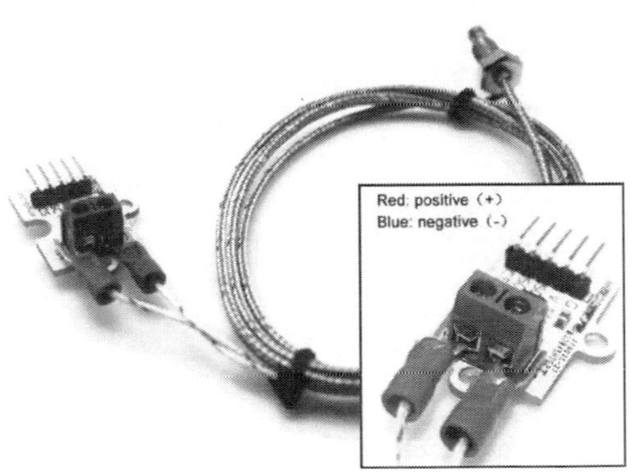

圖 17 MAX6675 K 型熱電偶感測器模組

表 7 MAX6675 K 型熱電偶感測器接腳表

	模組接腳	Arduino 開發板接腳	解說
max6675(PT100)	VCC	Arduino +5V	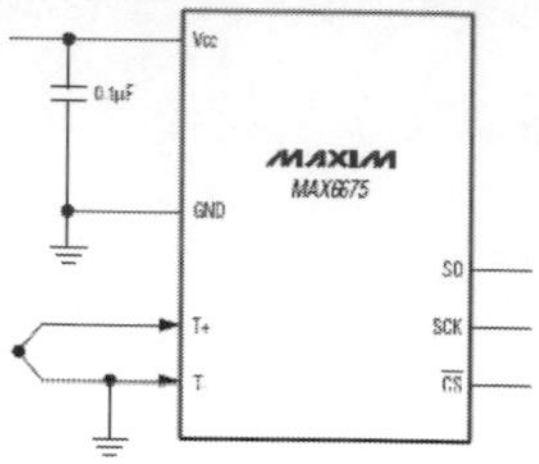
	GND	Arduino GND(共地接點)	
	SO(Signal Out)	Arduino Pin 8	
	CS(Chip Select)	Arduino Pin 9	
	SCK(System Clock(	Arduino Pin 10	
	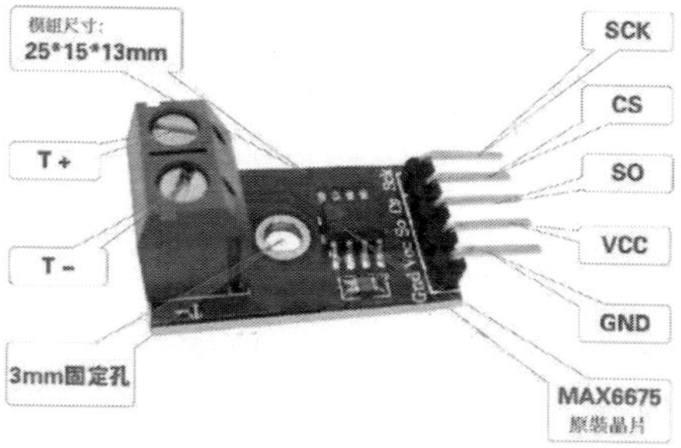		

完成 Arduino 開發板與 MAX6675 K 型熱電偶感測器模組連接之後，將下表所示**錯誤! 找不到參照來源。**之 max6675 溫度感測模組測試程式一鍵入 Arduino Sketch 之中，完成編譯後，上載到 Arduino 開發板進行測試，可以見到每隔一秒鐘(delay(2000))，可以看到讀取到溫度的資料。

表 8 max6675 溫度感測模組測試程式一

max6675 溫度感測模組測試程式一(max6675_test01)

<pre>#include <OneWire.h>
#include <DallasTemperature.h>
 	
// Arduino 數位腳位 2 接到 1-Wire 裝置
#define ONE_WIRE_BUS 2
 	
// 運用程式庫建立物件
OneWire oneWire(ONE_WIRE_BUS);
DallasTemperature sensors(&oneWire);</pre>

max6675 溫度感測模組測試程式一(max6675_test01)

```
void setup(void)
{
  Serial.begin(9600);
  Serial.println("Temperature Sensor");
  // 初始化
  sensors.begin();
}

void loop(void)
{
  // 要求匯流排上的所有感測器進行溫度轉換
  sensors.requestTemperatures();

  // 取得溫度讀數（攝氏）並輸出，
  // 參數 0 代表匯流排上第 0 個 1-Wire 裝置
  Serial.println(sensors.getTempCByIndex(0));

  delay(2000);
}
```

程式下載：https://github.com/brucetsao/eTemperature_Humidity

由下圖所示，可以看到透過 MAX6675 K 型熱電偶感測器模組可以讀取到外界溫度，並且該溫度式非常準確的溫度。

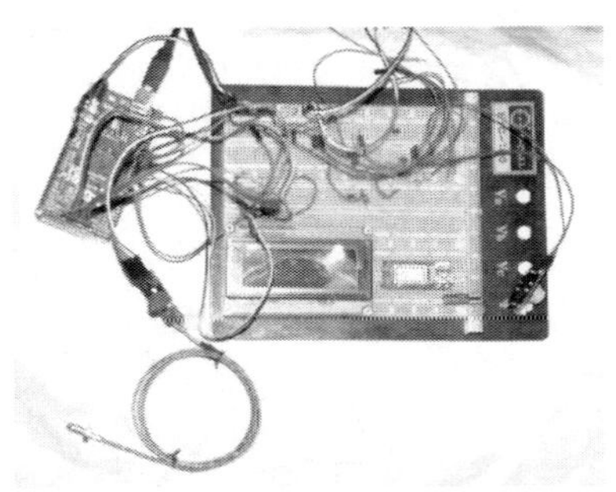

圖 18 max6675 溫度感測模組測試程式一畫面

章節小結

本章主要介紹之 Arduino 開發板使用與連接白金感溫電阻(PT100)溫度模組，透過本章節的解說，相信讀者會對連接、使用白金感溫電阻(PT100)溫度模組來量測溫度，有更深入的了解與體認。

5
CHAPTER

DHT11 溫濕度感測模組

如果我們要量測溫度，我們可以使用溫度感測器，如果我們又要量測濕度，我們可以使用量測感測器，這樣我們會需要很多的感測器，如下圖所示，所以本章節介紹最常使用，最便宜的溫濕度感測模組(DHT11)(曹永忠, 許智誠, et al., 2015a, 2015b, 2015c, 2015d; 曹永忠, 許智誠, & 蔡英德, 2015e, 2016c, 2016d; 曹永忠, 許碩芳, 許智誠, & 蔡英德, 2015a, 2015b)。

溫濕度感測模組(DHT11)

如果我們要量測溫度，我們可以使用溫度感測器，如果我們又要量測濕度，我們可以使用量測感測器，這樣我們會需要很多的感測器，所以本節介紹溫濕度感測模組(DHT11) (如下圖所示)，它主要是使用 DHT-11 作成溫濕度感測模組(DHT11)(曹永忠, 許智誠, et al., 2015e)。

圖 19 溫濕度感測模組(DHT11)

本實驗是採用溫濕度感測模組(DHT11)，如下圖所示，由於 DHT-11 溫濕度感測器需要搭配基本量測電路，所以我們使用溫濕度感測模組(DHT11)來當實驗主體，並不另外組立基本量測電路。

如下圖所示，先參考溫濕度感測模組(DHT11)腳位接法，在遵照下表之溫濕度感測模組(DHT11)接腳表進行電路組裝。

圖 20 溫濕度感測模組(DHT11)腳位圖

表 9 溫濕度感測模組(DHT11)接腳表

接腳	接腳說明	Arduino 開發板接腳
S	Vcc	電源 (+5V) Arduino +5V
2	GND	Arduino GND
3	Signal	Arduino digital Pin 7

接腳	接腳說明	接腳名稱
1	Ground (0V)	接地 (0V) Arduino GND
2	Supply voltage; 5V (4.7V – 5.3V)	電源 (+5V) Arduino +5V
3	SDA	Arduino SDA Pin
4	SCL	Arduino SCL Pin

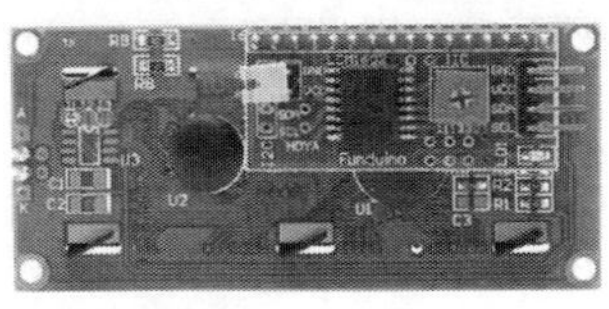

資料來源：Arduino 程式教學(常用模組篇):Arduino Programming (37 Sensor Modules)(曹永忠, 許智誠, et al., 2015b)

我們遵照前幾章所述，將 Arduino 開發板的驅動程式安裝好之後，我們打開 Arduino 開發板的開發工具：Sketch IDE 整合開發軟體，攥寫一段程式，如下表所示之溫濕度感測模組(DHT11)測試程式，我們就可以透過溫濕度感測模組(DHT11)來偵測任何溫度與濕度。

表 10 溫濕度感測模組測試程式

溫濕度感測模組測試程式(DHT11_sensor)

```
int DHPin=7;
byte dat[5];

byte read_data()
{
	byte data;
	for(int i=0; i<8;i++)
	{
		if(digitalRead(DHPin)==LOW)
			{

				while(digitalRead(DHPin)==LOW);				//等待
50us
					delayMicroseconds(30);
//判斷高電位的持續時間，以判定數據是 '0' 還是 '1'

				if(digitalRead(DHPin)==HIGH)
					data |=(1<<(7-i));
//高位在前，低位在後

				while(digitalRead(DHPin) == HIGH);				//數據
 '1' ，等待下一位的接收
			}
	}
	return data;
}

void start_test()
```

```
{
    digitalWrite(DHPin,LOW);                              //拉低總線，發開始
信號
    delay(30);                                                          //延
遲時間要大於 18ms，以便檢測器能檢測到開始訊號；
    digitalWrite(DHPin,HIGH);
    delayMicroseconds(40);                                   //等待感測器響應；
    PinMode(DHPin,INPUT);
  while(digitalRead(DHPin) == HIGH);
      delayMicroseconds(80);                                 //發出響應，拉低
总线 80us；
    if(digitalRead(DHPin) == LOW);
        delayMicroseconds(80);                                    //線路 80us 後
開始發送數據；

for(int i=0;i<4;i++)                                                  //接收溫溼度
數據，校验位不考慮；
    dat[i] = read_data();

      PinMode(DHPin,OUTPUT);
      digitalWrite(DHPin,HIGH);                                       //發送完數
據後釋放線路，等待下一次的開始訊號；
  }

void setup()
{
    Serial.begin(9600);
    PinMode(DHPin,OUTPUT);
}

void loop()
{
    start_test();
    Serial.print("Current humdity = ");
    Serial.print(dat[0], DEC);                                       //顯示濕度的
整數位；
    Serial.print('.');
    Serial.print(dat[1],DEC);                                         //顯示濕度
的小數位；
```

```
    Serial.println('%');
    Serial.print("Current temperature = ");
    Serial.print(dat[2], DEC);                              //顯示溫度的
整數位；
    Serial.print('.');
    Serial.print(dat[3],DEC);                               //顯示溫度的
小數位；
    Serial.println('C');
    delay(700);
  }
```

當然、如下圖所示，我們可以看到溫濕度感測模組測試程式結果畫面。

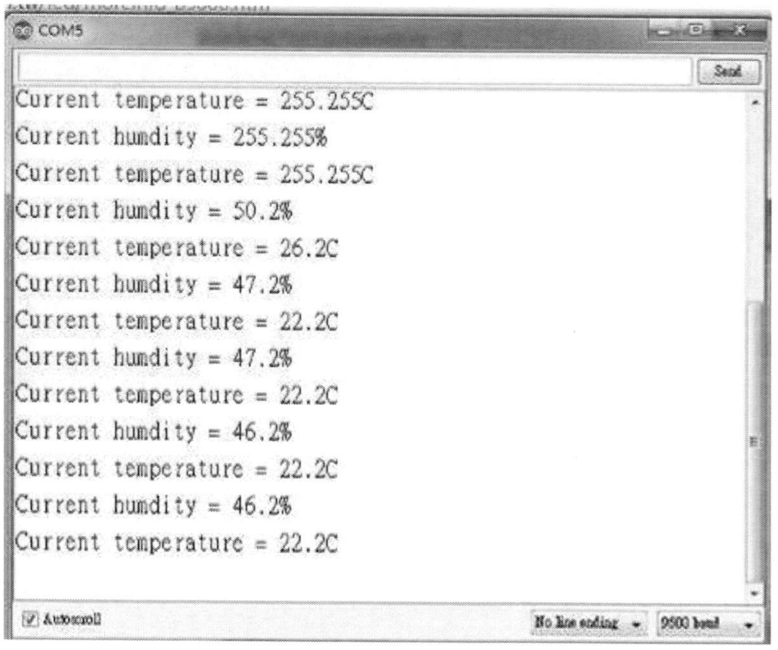

圖 21 溫濕度感測模組測試程式結果畫面

上面的程式我們並沒有使用 DHT11 的函式庫，所以整個程式變的很困難，也很難理解，所以作者寫了另外一版程式來使用 DHT11 的函式庫，使整個程式變的簡單、易學、易懂。

我們打開 Arduino 開發板的開發工具：Sketch IDE 整合開發軟體，攥寫一段程

式，如下表所示之 DHT11 溫濕度感測模組測試程式，我們就可以透過溫濕度感測模組(DHT11)來偵測任何溫度與濕度。

表 11 DHT11 溫濕度感測模組測試程式

DHT11 溫濕度感測模組測試程式(DHT11)

```
    int DHPin=7;
byte dat[5];

byte read_data()
{
      byte data;
      for(int i=0; i<8;i++)
      {
            if(digitalRead(DHPin)==LOW)
                  {

                        while(digitalRead(DHPin)==LOW);                              //等待
50us
                              delayMicroseconds(30);
//判斷高電位的持續時間，以判定數據是 '0' 還是 '1'

                        if(digitalRead(DHPin)==HIGH)
                              data |=(1<<(7-i));
//高位在前，低位在後

                        while(digitalRead(DHPin) == HIGH);                           //數據
 '1' ，等待下一位的接收
                  }
      }
      return data;
}

void start_test()
{
      digitalWrite(DHPin,LOW);                                         //拉低總線，發開始
信號
```

```
    delay(30);                                                          //延
遲時間要大於 18ms，以便檢測器能檢測到開始訊號；
    digitalWrite(DHPin,HIGH);
    delayMicroseconds(40);                                  //等待感測器響應；
    PinMode(DHPin,INPUT);
  while(digitalRead(DHPin) == HIGH);
      delayMicroseconds(80);                                  //發出響應，拉低
总线 80us；
    if(digitalRead(DHPin) == LOW);
        delayMicroseconds(80);                                   //線路 80us 後
開始發送數據；

for(int i=0;i<4;i++)                                                //接收溫溼度
數據，校验位不考慮；
    dat[i] = read_data();

      PinMode(DHPin,OUTPUT);
      digitalWrite(DHPin,HIGH);                                      //發送完數
據後釋放線路，等待下一次的開始訊號；
  }

void setup()
{
    Serial.begin(9600);
    PinMode(DHPin,OUTPUT);
}

void loop()
{
    start_test();
    Serial.print("Current humdity = ");
    Serial.print(dat[0], DEC);                                        //顯示濕度的
整數位；
    Serial.print('.');
    Serial.print(dat[1],DEC);                                         //顯示濕度
的小數位；
    Serial.println('%');
    Serial.print("Current temperature = ");
    Serial.print(dat[2], DEC);                                        //顯示溫度的
```

```
整數位；
      Serial.print('.');
      Serial.print(dat[3],DEC);                                    //顯示溫度的
小數位；
      Serial.println('C');
      delay(700);
   }
```

程式下載：https://github.com/brucetsao/eTemperature_Humidity

當然、如下圖所示，我們可以看到溫濕度感測模組測試程式結果畫面。

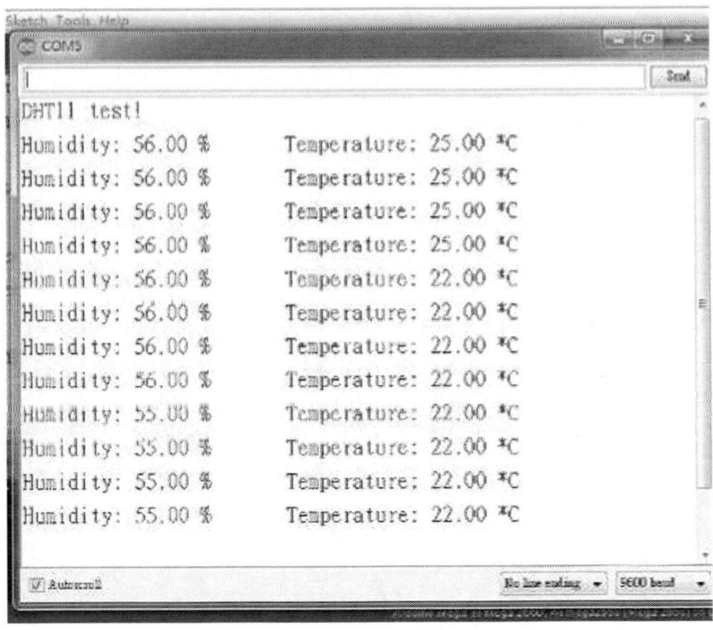

圖 22 DHT11 溫濕度感測模組測試程式結果畫面

章節小結

本章節內容主要是解釋如何使用溫溼度感測器的功能，所以我們必須了解目前常用的溫溼度感測器使用方法，方能繼續往下實作，繼續進行我們的實驗。

6 CHAPTER

DHT21/22 溫濕度感測模組器

如果我們要量測溫度，我們可以使用溫度感測器，如果我們又要量測濕度，我們可以使用量測感測器，這樣我們會需要很多的感測器，如下圖所示，所以本章節介紹最常使用，更專業的溫濕度感測模組(DHT21/22)(曹永忠, 許智誠, et al., 2015a, 2015b, 2015c, 2015d, 2015e; 曹永忠 et al., 2016c, 2016d; 曹永忠, 許碩芳, et al., 2015a, 2015b)。

溫濕度感測模組(DHT21/22)

本實驗為了讓 Arduino 開發板進階使用，使用了更進階的的 DHT22 溫濕度感測模組(如下圖所示)(曹永忠, 許智誠, et al., 2015e)，本模組只要將 Vcc 接到 Arduino 開發板+5V 腳位，Gnd 接到 Arduino 開發板 Gnd 腳位，DAT 接到 Arduino 開發板 Digital Input 腳位 2，對於 Arduino 開發板外部插斷接腳不太了解的讀者，可以參閱 http://arduino.cc/en/Reference/AttachInterrupt)，再執行下列程式。

圖 23　DHT22 溫濕度感測模組

表 12 DHT22 溫濕度感測模組接腳圖

DHT22 溫濕度感測模組	Arduino 開發板接腳	解說
DAT	Arduino digital Input pin 2	DHT22 資料輸出腳位
5V	Arduino pin 5V	5V 陽極接點
GND	Arduino pin Gnd	共地接點

其餘關於 DHT22 溫濕度感測器的細部資料，本文使用的 DHT-22 函式庫，是採用網路上 Seeed-Studio 針對 DHT22 Temperature and Humidity Sensor 所攥寫的 Arduino library，Seeed-Studio 在 GitHub, Inc.撰寫之程式碼，其下載網址為：https://github.com/Seeed-Studio/Grove_Temperature_And_Humidity_Sensor，特此感謝分享。

我們將下表之 DHT22 溫濕度感測器讀取溫濕度測試程式攥寫好之後，編譯完成後上傳到 Arduino 開發板，

表 13 DHT22 溫濕度感測器讀取溫濕度測試程式

DHT22 溫濕度感測器讀取溫濕度測試程式(DHT_22_test1)

```
// Example testing sketch for various DHT humidity/temperature sensors
// Written by ladyada, public domain

#include "DHT.h"

#define DHTPIN 2 // what pin we're connected to

// Uncomment whatever type you're using!
//#define DHTTYPE DHT11 // DHT 11
#define DHTTYPE DHT22 // DHT 22 (AM2302)
//#define DHTTYPE DHT21 // DHT 21 (AM2301)

// Connect pin 1 (on the left) of the sensor to +5V
// Connect pin 2 of the sensor to whatever your DHTPIN is
// Connect pin 4 (on the right) of the sensor to GROUND
```

DHT22 溫濕度感測器讀取溫濕度測試程式(DHT_22_test1)

```
// Connect a 10K resistor from pin 2 (data) to pin 1 (power) of the sensor

DHT dht(DHTPIN, DHTTYPE);

void setup()
{
        Serial.begin(9600);
        Serial.println("DHTxx test!");

        dht.begin();
}

void loop()
{
        // Reading temperature or humidity takes about 250 milliseconds!
        // Sensor readings may also be up to A0 seconds 'old' (its a very slow sensor)
        float h = dht.readHumidity();
        float t = dht.readTemperature();

        // check if returns are valid, if they are NaN (not a number) then something
went wrong!
        if (isnan(t) || isnan(h))
        {
                Serial.println("Failed to read from DHT");
        }
        else
        {
                Serial.print("Humidity: ");
                Serial.print(h);
                Serial.print(" %\t");
                Serial.print("Temperature: ");
                Serial.print(t);
                Serial.println(" *C");
        }
}
```

參考資料來源：Grove- Temperature and Humidity Sensor

(http://www.seeedstudio.com/wiki/Grove-_Temperature_and_Humidity_Sensor)

程式下載：https://github.com/brucetsao/eTemperature_Humidity

上述程式執行後，可以見到下圖之 DHT22 溫濕度感測器讀取溫濕度測試程式畫面結果，也可以輕易讀到外界的溫度與濕度了。

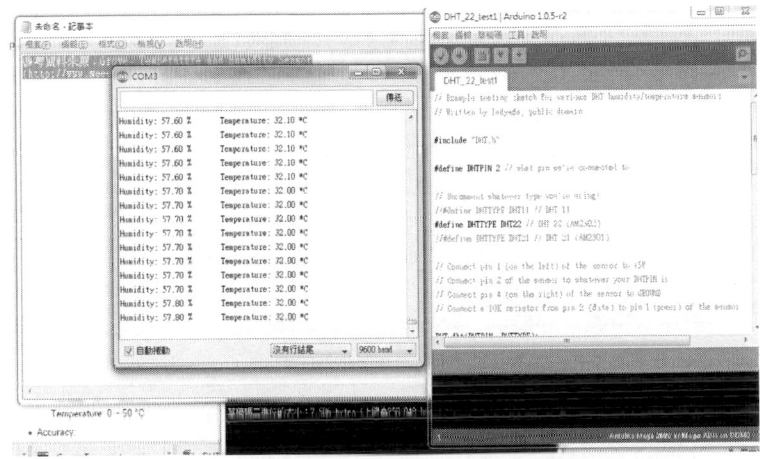

圖 24 DHT22 溫濕度感測器讀取溫濕度測試程式畫面結果

章節小結

本章節內容主要是解釋如何使用溫溼度感測器的功能，所以我們必須了解目前常用的溫溼度感測器使用方法，方能繼續往下實作，繼續進行我們的實驗。

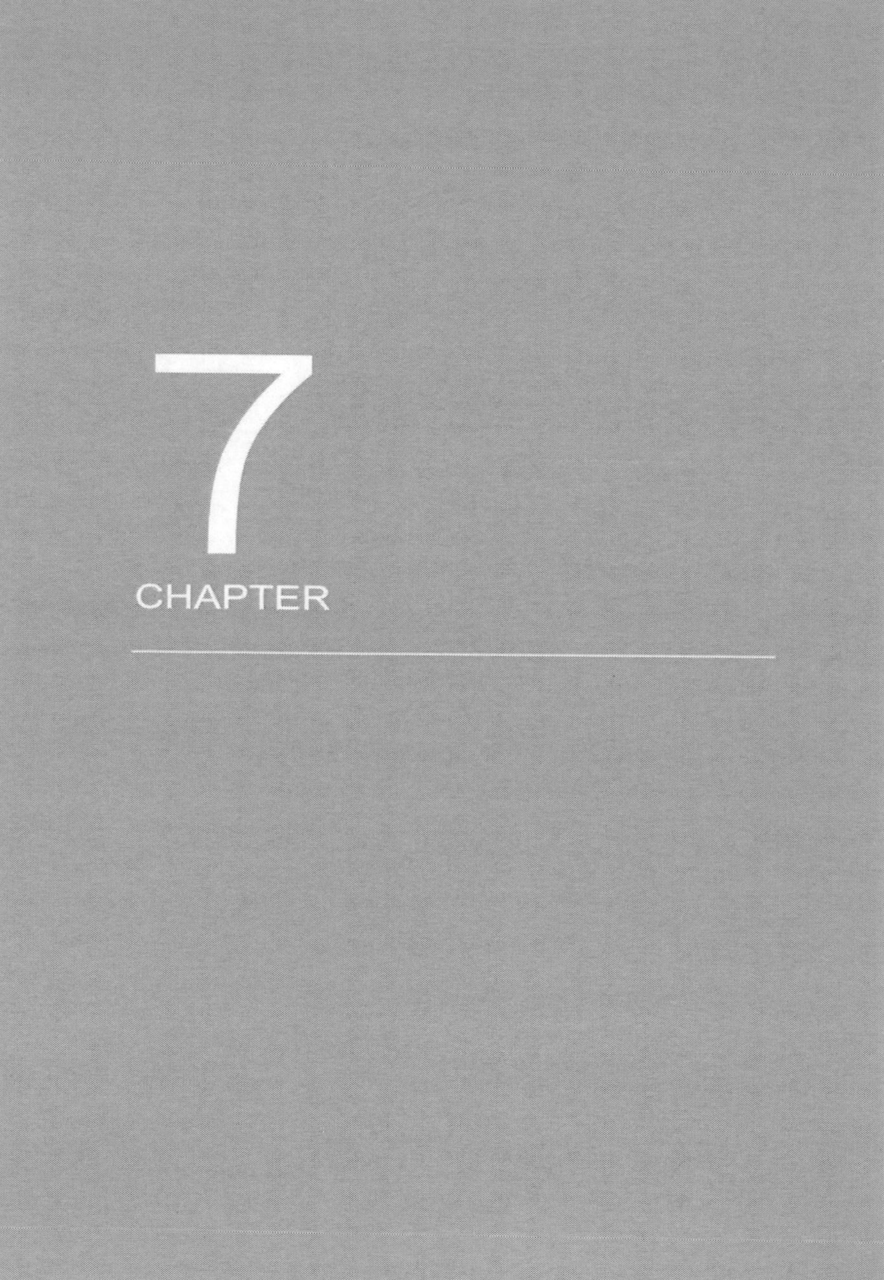
7
CHAPTER

SHT31 溫濕度感測模組

如果我們要量測溫度，我們可以使用溫度感測器，如果我們又要量測濕度，我們可以使用量測感測器，這樣我們會需要很多的感測器，如下圖所示，所以本章節介紹專業型溫濕度感測模組(SHT31)(曹永忠, 許智誠, & 蔡英德, 2016a, 2016b)。

溫濕度感測模組(SHT31)

本實驗為了讓 Arduino 開發板進階使用，使用了工業級的 SHT31 溫濕度感測模組(如下圖所示)，本模組只要將 Vcc 接到 Arduino 開發板+5V 腳位，Gnd 接到 Arduino 開發板 Gnd 腳位，其 I2C 腳位接到 Arduino 開發板 I2C 腳位，就可以完成電子線路。

圖 25　SHT31 溫濕度感測模組

讀者可以參考下圖所示之 SHT31 溫濕度感測連接電路圖，對於 I2C 之腳位不太明瞭的，也可以參考下表之腳位說明，進行電路組立(曹永忠 et al., 2016a, 2016b)。

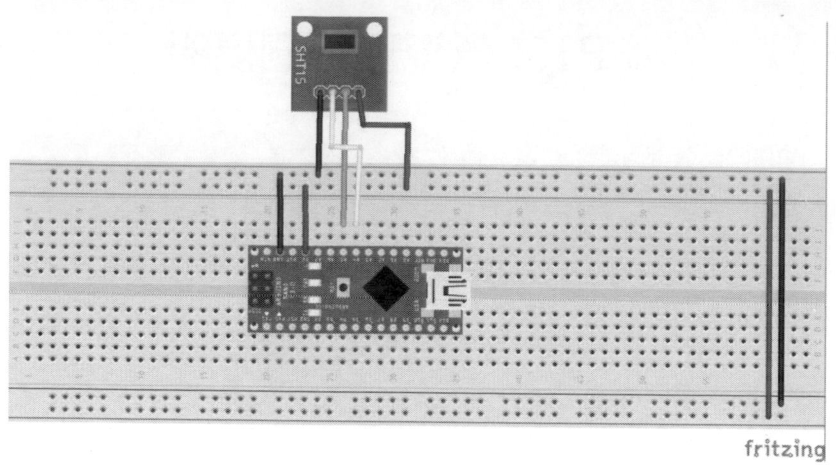

圖 26　SHT31 溫濕度感測連接電路圖

讀者如果對於 I2C 腳位不太了解的，可以先參考下表，或參閱筆者拙作『Arduino 程式教學(基本語法篇):Arduino Programming (Language & Syntax)』(曹永忠 et al., 2016c, 2016d)、『Arduino 程式教學(入門篇):Arduino Programming (Basic Skills & Tricks)』(曹永忠, 許智誠, et al., 2015a, 2015d)、『Arduino 程式教學(常用模組篇):Arduino Programming (37 Sensor Modules)』(曹永忠, 許智誠, et al., 2015b, 2015c)、了解不同開發板，期 I2C 腳位差異。

讀者可以參考下表之腳位說明進行電路組立。

表 14 SHT31 溫濕度感測模組接腳表

SHT31 溫濕度感測模組	Arduino 開發板接腳	解說
SHT31 SDA	SDA/A4	I2C SDA 腳位
SHT31 SCL	SCL/A5	I2C SCL 腳位
SHT31 5V	Arduino pin 5V	5V 陽極接點
SHT31 GND	Arduino pin Gnd	共地接點

其餘關於 SHT31 溫濕度感測器的細部資料，本文使用的 SHT31 函式庫，是採

用 Adafruit 公司出廠的 Adafruit Sensiron SHT31-D Temperature & Humidity Sensor Breakout，網址如下：https://www.adafruit.com/products/2857，搭配產品所攥寫的 Arduino library，其下載網址為：https://github.com/adafruit/Adafruit_SHT31，特此感謝分享。

也可以到筆者的 Github(https://github.com/brucetsao)，其網址：https://github.com/brucetsao/LIB_for_MCU，有各種開發板可以使用的函式庫，讀者可以去下載所需要的函式庫。在安裝好函示庫之後，我們將下表之 SHT31 溫濕度感測器讀取溫濕度測試程式攥寫好之後，編譯完成後上傳到 Arduino 開發板。

表 15 SHT31 溫濕度感測器讀取溫濕度測試程式

SHT31 溫濕度感測器讀取溫濕度測試程式(SHT31test)

```
/***************************************************
  This is an example for the SHT31-D Humidity & Temp Sensor

  Designed specifically to work with the SHT31-D sensor from Adafruit
  ----> https://www.adafruit.com/products/2857

  These sensors use I2C to communicate, 2 pins are required to
  interface
 ****************************************************/

#include <Arduino.h>
#include <Wire.h>
#include "Adafruit_SHT31.h"

Adafruit_SHT31 sht31 = Adafruit_SHT31();

#if defined(ARDUINO_ARCH_SAMD)
// for Zero, output on USB Serial console, remove line below if using programming port
to program the Zero!
   #define Serial SerialUSB
#endif

void setup() {
```

SHT31 溫濕度感測器讀取溫濕度測試程式(SHT31test)

```
#ifndef ESP8266
  while (!Serial);       // will pause Zero, Leonardo, etc until serial console opens
#endif
  Serial.begin(9600);
  Serial.println("SHT31 test");
  if (! sht31.begin(0x44))
  {    // Set to 0x45 for alternate i2c addr
    Serial.println("Couldn't find SHT31");
    while (1) delay(1);
  }
}


void loop() {
  float t = sht31.readTemperature();
  float h = sht31.readHumidity();

  if (! isnan(t)) {   // check if 'is not a number'
    Serial.print("Temp *C = "); Serial.println(t);
  } else {
    Serial.println("Failed to read temperature");
  }

  if (! isnan(h)) {   // check if 'is not a number'
    Serial.print("Hum. % = "); Serial.println(h);
  } else {
    Serial.println("Failed to read humidity");
  }
  Serial.println();
  delay(1000);
    }
```

參考資料來源：Adafruit Sensiron SHT31-D Temperature & Humidity Sensor Breakout Wiring & Test (https://learn.adafruit.com/adafruit-sht31-d-temperature-and-humidity-sensor-breakout/wiring-and-test)

程式下載：https://github.com/brucetsao/eTemperature_Humidity

上述程式執行後，可以見到下圖之 SHT31 溫濕度感測器讀取溫濕度測試程式畫面結果，也可以輕易讀到外界的溫度與濕度了。

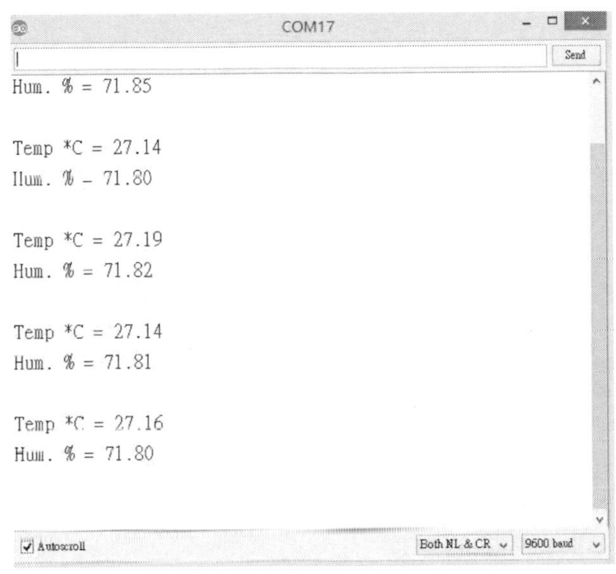

圖 27 SHT31 溫濕度感測器讀取溫濕度測試程式畫面結果

章節小結

本章節內容主要是解釋如何使用溫溼度感測器的功能，所以我們必須了解目前常用的溫溼度感測器使用方法，方能繼續往下實作，繼續進行我們的實驗。

CHAPTER

SH21 溫濕度感測模組

如果我們要量測溫度，我們可以使用溫度感測器，如果我們又要量測濕度，我們可以使用量測感測器，這樣我們會需要很多的感測器，如下圖所示，所以本章節介紹專業型溫濕度感測模組(SHT21)。

溫濕度感測模組(SHT21)

SHT21 溫濕度感測模組是很常用且易用的溫溼度感測器元件，在元器件的應用上也只需要一個 SHT21 元件，只利用一個 I2C 通訊介面就可以，將讀取的溫度、溼度轉換為實際的溫溼度，透過 I2C 通訊回傳到開發版。

所需的元器件如下。

- 溫濕度感測模組
- 麵包板*1
- 麵包板跳線*1 紮

如下圖所示，這個實驗我們需要用到的實驗硬體有下圖.(a)的 Arduino Mega 2560 與下圖.(b) USB 下載線、下圖.(c) SHT21 溫濕度感測模組、下圖.(d).LCD1602 液晶顯示器：

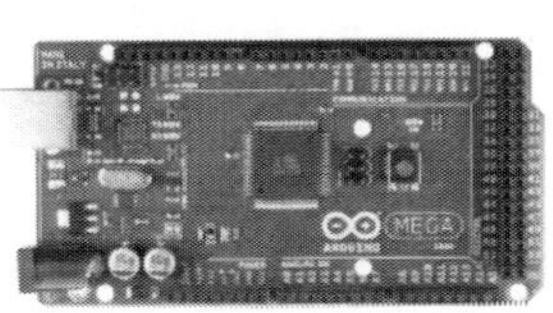

(a).Arduino Mega 2560

(b). USB 下載線

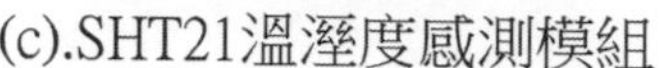

(c).SHT21溫溼度感測模組

(d).LCD1602液晶顯示器

圖 28 SHT21 溫濕度感測模組所需材料表

我們遵照前幾章所述，將 Arduino 開發板的驅動程式安裝好之後，我們打開 Arduino 開發板的開發工具：Sketch IDE 整合開發軟體(軟體下載請到：https://www.arduino.cc/en/Main/Software)，編寫一段程式，如下表所示之 SHT21 溫溼度感測模組程式程式，讓 Arduino 讀取 SHT21 溫溼度感測模組程式，並把溫度顯示在 Sketch 的監控畫面。

本實驗為了讓 Arduino 開發板進階使用，使用了工業級的 SHT21 溫濕度感測模組(如下圖所示)，本模組只要將 Vcc 接到 Arduino 開發板+5V 腳位，Gnd 接到 Arduino 開發板 Gnd 腳位，其 I2C 腳位接到 Arduino 開發板 I2C 腳位，就可以完成電子線路。

圖 29　SHT21 溫濕度感測模組腳位

讀者可以參考下圖所示之 SHT21 溫濕度感測模組連接電路圖，對於 I2C 之腳位不太明瞭的，也可以參考下表之腳位說明，進行電路組立(曹永忠 et al., 2016a, 2016b)。

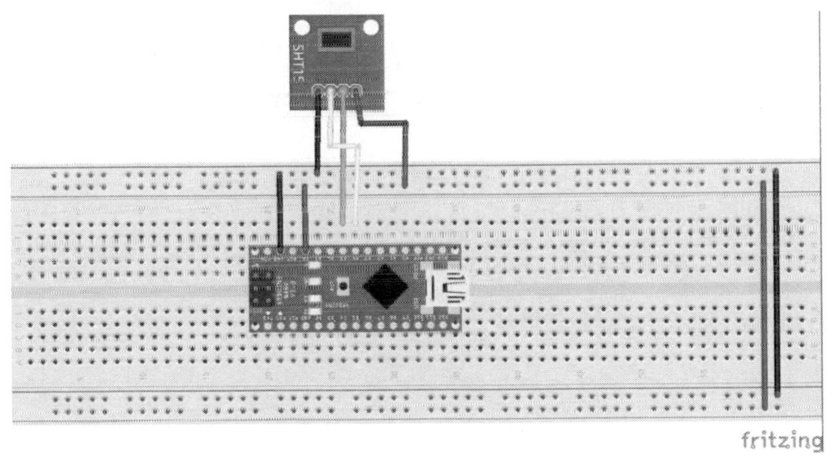

圖 30　SHT21 溫濕度感測連接電路圖

讀者如果對於 I2C 腳位不太了解的，可以先參考下表，或參閱筆者拙作『Arduino 程式教學(基本語法篇):Arduino Programming (Language & Syntax)』(曹永忠 et al., 2016c, 2016d)、『Arduino 程式教學(入門篇):Arduino Programming (Basic Skills & Tricks)』(曹永忠, 許智誠, et al., 2015a, 2015d)、『Arduino 程式教學(常用模組篇):Arduino Programming (37 Sensor Modules)』(曹永忠, 許智誠, et al., 2015b, 2015c)、

了解不同開發板，期 I2C 腳位差異。

讀者可以參考下表之腳位說明進行電路組立。

表 16 SHT21 溫濕度感測模組接腳表

SHT31 溫濕度感測模組	Arduino 開發板接腳	解說
SHT21 SDA	SDA/A4(UNO)	I2C SDA 腳位
SHT21 SCL	SCL/A5(UNO)	I2C SCL 腳位
SHT21 5V	Arduino pin 5V	5V 陽極接點
SHT21 GND	Arduino pin Gnd	共地接點

其餘關於 SHT21 溫濕度感測器的細部資料，本文使用的 SHT21 函式庫，是採用 Misenso Electronics(https://github.com/misenso) 的函式庫，其下載網址為：https://github.com/misenso/SHT2x-Arduino-Library，特此感謝分享。

也可以到筆者的 Github(https://github.com/brucetsao)，其網址：https://github.com/brucetsao/LIB_for_MCU，有各種開發板可以使用的函式庫，讀者可以去下載所需要的函式庫。在安裝好函示庫之後，我們將下表之 SHT21 溫濕度感測器讀取溫濕度測試程式攥寫好之後，編譯完成後上傳到 Arduino 開發板，

表 17 SHT21 溫溼度感測模組讀取溫濕度測試程式

SHT21 溫溼度感測模組讀取溫濕度測試程式(ReadSHT2x)

```
/****************************************************************
 * ReadSHT2x
 *  An example sketch that reads the sensor and prints the
 *  relative humidity to the PC's serial port
 *
 *  Tested with:
 *    - SHT21-Breakout Humidity sensor from Modern Device
```

SHT21 溫溼度感測模組讀取溫濕度測試程式(ReadSHT2x)

```
 *       - SHT2x-Breakout Humidity sensor from MisensO Electronics
 ******************************************************************/

#include <Wire.h>
#include <SHT2x.h>


void setup()
{
  Wire.begin();
  Serial.begin(9600);
}

void loop()
{
  Serial.print("Humidity(%RH): ");
  Serial.print(SHT2x.GetHumidity());
  Serial.print("        Temperature(C): ");
  Serial.println(SHT2x.GetTemperature());

  delay(1000);
    }
```

參考資料來源：https://github.com/misenso/SHT2x-Arduino-Library

程式下載：https://github.com/brucetsao/eTemperature_Humidity

上述程式執行後，可以見到下圖之 SHT21 溫濕度感測器讀取溫濕度測試程式畫面結果，也可以輕易讀到外界的溫度與濕度了。

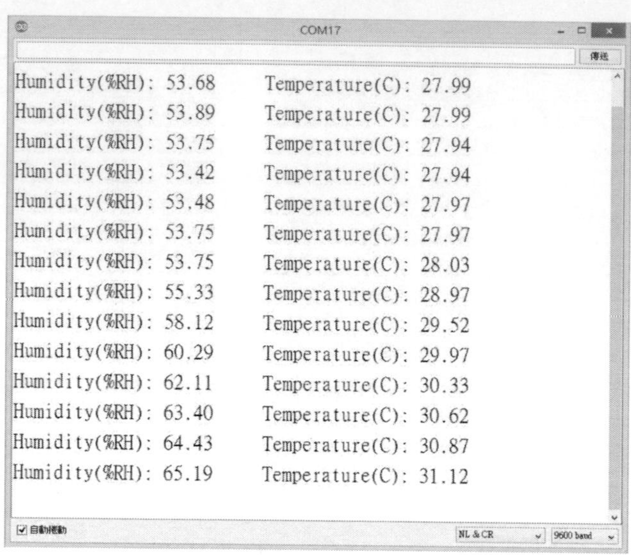

圖 31 SHT21 溫濕度感測模組讀取溫濕度測試程式畫面結果

章節小結

本章節內容主要是解釋如何使用 SHT21 溫濕度感測模組的功能，所以我們必須了解目前常用的溫溼度感測器使用方法，方能繼續往下實作，繼續進行我們的實驗。

CHAPTER

HTU21D 溫濕度感測模組

如果我們要量測溫度，我們可以使用溫度感測器，如果我們又要量測濕度，我們可以使用量測感測器，這樣我們會需要很多的感測器，如下圖所示，所以本章節介紹專業型溫濕度感測模組(HTU21D)(曹永忠 et al., 2016a, 2016b)。

溫濕度感測模組(HTU21D)

溫濕度感測模組是很常用且易用的溫溼度感測器元件，在元器件的應用上也只需要一個 HTU21D 元件，只利用一個 I2C 通訊介面就可以，將讀取的溫度、溼度轉換為實際的溫溼度，透過 I2C 通訊回傳到開發版。

所需的元器件如下。

- HTU21D 溫濕度感測模組
- 麵包板*1
- 麵包板跳線*1 紮

如下圖所示，這個實驗我們需要用到的實驗硬體有下圖.(a)的 Arduino Mega 2560 與下圖.(b) USB 下載線、下圖.(c) HTU21D 溫度感測器、下圖.(d).LCD1602 液晶顯示器：

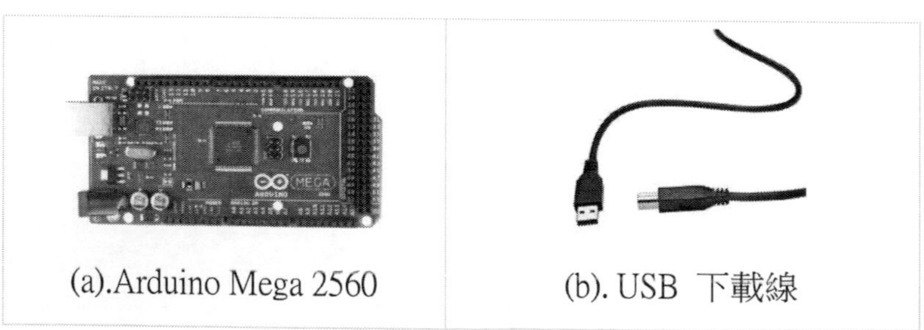

(a).Arduino Mega 2560 　(b). USB 下載線

(c). HTU21D溫溼度感測模組

(d).LCD1602液晶顯示器

圖 32 HTU21D 溫濕度感測模組所需材料表

我們遵照前幾章所述，將 Arduino 開發板的驅動程式安裝好之後，我們打開 Arduino 開發板的開發工具：Sketch IDE 整合開發軟體(軟體下載請到：https://www.arduino.cc/en/Main/Software)，編寫一段程式，如下表所示之 HTU21D 溫溼度感測模組程式程式，讓 Arduino 讀取 HTU21D 溫溼度感測模組程式，並把溫度顯示在 Sketch 的監控畫面。

本實驗為了讓 Arduino 開發板進階使用，使用了工業級的 HTU21D 溫濕度感測模組(如下圖所示)，本模組只要將 Vcc 接到 Arduino 開發板+5V 腳位，Gnd 接到 Arduino 開發板 Gnd 腳位，其 I2C 腳位接到 Arduino 開發板 I2C 腳位，就可以完成電子線路。

圖 33　HTU21D 溫濕度感測模組腳位

讀者可以參考下圖所示之 HTU21D 溫濕度感測模組連接電路圖，對於 I2C 之腳位不太明瞭的，也可以參考下表之腳位說明，進行電路組立(曹永忠 et al., 2016a,

2016b)。

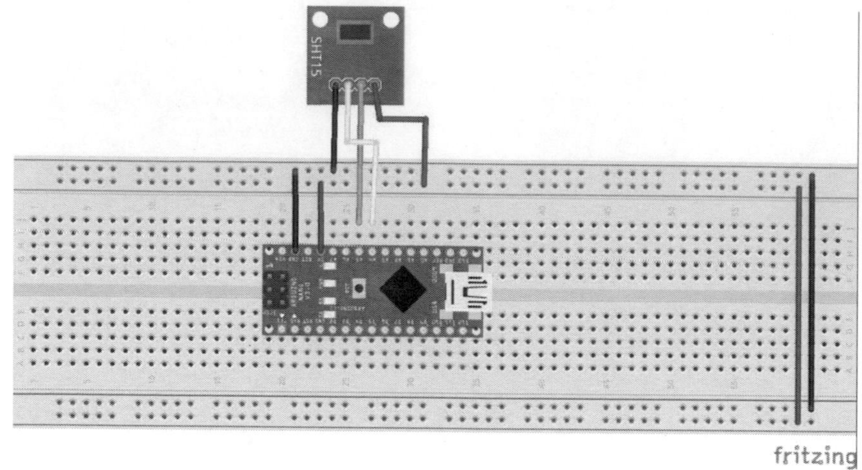

圖 34　HTU21D 溫濕度感測連接電路圖

讀者如果對於 I2C 腳位不太了解的，可以先參考下表，或參閱筆者拙作『Arduino 程式教學(基本語法篇):Arduino Programming (Language & Syntax)』(曹永忠 et al., 2016c, 2016d)、『Arduino 程式教學(入門篇):Arduino Programming (Basic Skills & Tricks)』(曹永忠, 許智誠, et al., 2015a, 2015d)、『Arduino 程式教學(常用模組篇):Arduino Programming (37 Sensor Modules)』(曹永忠, 許智誠, et al., 2015b, 2015c)、了解不同開發板，期 I2C 腳位差異。

讀者可以參考下表之腳位說明進行電路組立。

表 18 HTU21D 溫濕度感測模組接腳表

SHT31 溫濕度感測模組	Arduino 開發板接腳	解說
SHT21 SDA(DA)	SDA/A4(UNO)	I2C SDA 腳位
SHT21 SCL(CL)	SCL/A5(UNO)	I2C SCL 腳位
SHT21 5V	Arduino pin 5V	5V 陽極接點
SHT21 GND	Arduino pin Gnd	共地接點

其餘關於 HTU21D 溫濕度感測器的細部資料，本文使用的 HTU21D 函式庫，是採用 SparkFun Electronics (https://www.sparkfun.com/)的函式庫，其下載網址為：https://github.com/sparkfun/SparkFun_HTU21D_Breakout_Arduino_Library/archive/master.zip，特此感謝分享。

也可以到筆者的 Github(https://github.com/brucetsao)，其網址：https://github.com/brucetsao/LIB_for_MCU，有各種開發板可以使用的函式庫，讀者可以去下載所需要的函式庫。在安裝好函示庫之後，我們將下表之 SHT21 溫濕度感測器讀取溫濕度測試程式攥寫好之後，編譯完成後上傳到 Arduino 開發板，

表 19 HTU21D 溫溼度感測模組測試程式

HTU21D 溫溼度感測模組測試程式(SparkFun_HTU21D_Demo)
/* HTU21D Humidity Sensor Example Code By: Nathan Seidle SparkFun Electronics Date: September 15th, 2013 License: This code is public domain but you buy me a beer if you use this and we meet someday (Beerware license).   Uses the HTU21D library to display the current humidity and temperature   Open serial monitor at 9600 baud to see readings. Errors 998 if not sensor is detected. Error 999 if CRC is bad.   Hardware Connections (Breakoutboard to Arduino): -VCC = 3.3V -GND = GND -SDA = A4 (use inline 330 ohm resistor if your board is 5V) -SCL = A5 (use inline 330 ohm resistor if your board is 5V)   */

HTU21D 溫溼度感測模組測試程式(SparkFun_HTU21D_Demo)

```
#include <Wire.h>
#include "SparkFunHTU21D.h"

//Create an instance of the object
HTU21D myHumidity;

void setup()
{
  Serial.begin(9600);
  Serial.println("HTU21D Example!");

  myHumidity.begin();
}

void loop()
{
  float humd = myHumidity.readHumidity();
  float temp = myHumidity.readTemperature();

  Serial.print("Time:");
  Serial.print(millis());
  Serial.print(" Temperature:");
  Serial.print(temp, 1);
  Serial.print("C");
  Serial.print(" Humidity:");
  Serial.print(humd, 1);
  Serial.print("%");

  Serial.println();
  delay(1000);
    }
```

參考資料來源：

https://learn.sparkfun.com/tutorials/htu21d-humidity-sensor-hookup-guide/htu21d-library-and-functions

程式下載：https://github.com/brucetsao/eTemperature_Humidity

上述程式執行後，可以見到下圖之 HTU21D 溫濕度感測器讀取溫濕度測試程式畫面結果，也可以輕易讀到外界的溫度與濕度了。

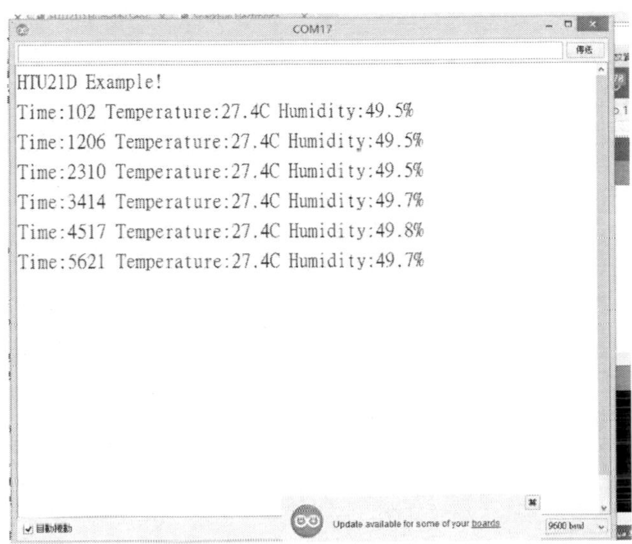

圖 35 HTU21D 溫濕度感測模組測試程式畫面結果

章節小結

本章節內容主要是解釋如何使用溫溼度感測器的功能，所以我們必須了解目前常用的溫溼度感測器使用方法，方能繼續往下實作，繼續進行我們的實驗。

本書總結

筆者對於 Arduino 相關的書籍，也出版許多書籍，感謝許多有心的讀者提供筆者許多寶貴的意見與建議，筆者群不勝感激，許多讀者希望筆者可以推出更多的入門書籍給更多想要進入『Arduino』、『Maker』這個未來大趨勢，所有才有這個入門系列的產生。

本系列叢書的特色是一步一步教導大家使用更基礎的東西，來累積各位的基礎能力，讓大家能更在 Maker 自造者運動中，可以拔的頭籌，所以本系列是一個永不結束的系列，只要更多的東西被製造出來，相信筆者會更衷心的希望與各位永遠在這條 Maker 路上與大家同行。

作者介紹

曹永忠 (Yung-Chung Tsao) ，目前為自由作家暨專業 Maker，專研於軟體工程、軟體開發與設計、物件導向程式設計，商品攝影及人像攝影。長期投入創客運動、資訊系統設計與開發、企業應用系統開發、軟體工程、新產品開發管理、商品及人像攝影等領域，並持續發表作品及相關專業著作。

Email:prgbruce@gmail.com

Line ID：dr.brucetsao

部落格：http://taiwanarduino.blogspot.tw/

臉書社群(Arduino.Taiwan)：

https://www.facebook.com/groups/Arduino.Taiwan/

Youtube：

https://www.youtube.com/channel/UCcYG2yY_u0m1aotcA4hrRgQ

許智誠 (Chih-Cheng Hsu)，美國加州大學洛杉磯分校(UCLA) 資訊工程系博士，曾任職於美國 IBM 等軟體公司多年，現任教於中央大學資訊管理學系專任副教授，主要研究為軟體工程、設計流程與自動化、數位教學、雲端裝置、多層式網頁系統、系統整合。

Email: khsu@mgt.ncu.edu.tw

蔡英德 (Yin-Te Tsai)，國立清華大學資訊科學系博士，目前是靜宜大學資訊傳播工程學系教授、靜宜大學計算機及通訊中心主任，主要研究為演算法設計與分析、生物資訊、軟體開發、視障輔具設計與開發。

Email:yttsai@pu.edu.tw

附 錄

DallasTemperature 函數用法

Arduino 開發版驅動 DS18B20 溫度感測模組，需要 DallasTemperature 函數庫，而 DallasTemperature 函數庫則需要 OneWire 函數庫，讀者可以在本書附錄中找到這些函市庫，也可以到作者 Github(https://github.com/brucetsao)網站中，在本書原始碼目錄 https://github.com/brucetsao/libraries，下載到 DallasTemperature、OneWire 等函數庫。

下列簡單介紹 DallasTemperature 函式庫內各個函式市的解釋與用法:

- uint8_t getDeviceCount(void)，回傳 1-Wire 匯流排上有多少個裝置。
- typedef uint8_t DeviceAddress[8]，裝置的位址。
- bool getAddress(uint8_t*, const uint8_t)，回傳某個裝置的位址。
- uint8_t getResolution(uint8_t*)，取得某裝置的溫度解析度（9~12 bits，分別對應 0.5℃、0.25℃、0.125℃、0.0625℃)，參數為位址。
- bool setResolution(uint8_t*, uint8_t)，設定某裝置的溫度解析度。
- bool requestTemperaturesByAddress(uint8_t*)，命令某感測器進行溫度轉換，參數為位址。
- bool requestTemperaturesByIndex(uint8_t)，同上，參數為索引值。
- float getTempC(uint8_t*)，取得溫度讀數，參數為位址。
- float getTempCByIndex(uint8_t)，取得溫度讀數，參數為索引值。
- 另有兩個靜態成員函式可作攝氏華氏轉換。
  - static float toFahrenheit(const float)
  - static float toCelsius(const float)

DHT11 Datasheet

数字温湿度传感器

DHT11

- ▶相对湿度和温度测量
- ▶全部校准，数字输出
- ▶卓越的长期稳定性
- ▶无需额外部件
- ▶超长的信号传输距离
- ▶超低能耗
- ▶4 引脚安装
- ▶完全互换

DHT11产品概述

DHT11数字温湿度传感器是一款含有已校准数字信号输出的温湿度复合传感器。它应用专用的数字模块采集技术和温湿度传感技术，确保产品具有极高的可靠性与卓越的长期稳定性。传感器包括一个电阻式感湿元件和一个NTC测温元件，并与一个高性能8位单片机相连接。因此该产品具有品质卓越、超快响应、抗干扰能力强、性价比极高等优点。每个DHT11传感器都在极为精确的湿度校验室中进行校准。校准系数以程序的形式储存在OTP内存中，传感器内部在检测信号的处理过程中要调用这些校准系数。单线制串行接口，使系统集成变得简易快捷。超小的体积、极低的功耗，信号传输距离可达20米以上，使其成为各类应用甚至最为苛刻的应用场合的最佳选则。产品为 4 针单排引脚封装。连接方便，特殊封装形式可根据用户需求而提供。

应用领域

- ▶暖通空调
- ▶汽车
- ▶消费品
- ▶气象站
- ▶湿度调节器
- ▶除湿器
- ▶测试及检测设备
- ▶数据记录器
- ▶自动控制
- ▶家电
- ▶医疗

订货信息

型号	测量范围	测湿精度	测温精度	分辨力	封装
DHT11	20－90%RH 0－50℃	±5%RH	±2℃	1	4针单排直插

1、传感器性能说明

参数	条件	Min	Typ	Max	单位
湿度					
分辨率		1	1	1	%RH
			8		Bit
重复性			±1		%RH
精度	25℃		±4		%RH
	0—50℃			±5	%RH
互换性	可完全互换				
量程范围	0℃	30		90	%RH
	25℃	20		90	%RH
	50℃	20		80	%RH
响应时间	1/e(63%)25℃, 1m/s 空气	6	10	15	S
迟滞			±1		%RH
长期稳定性	典型值		±1		%RH/yr
温度					
分辨率		1	1	1	℃
		8	8	8	Bit
重复性			±1		℃
精度		±1		±2	℃
量程范围		0		50	℃
响应时间	1/e(63%)	6		30	S

2、 接口说明

建议连接线长度短于20米时用5K上拉电阻，大于20米时根据实际情况使用合适的上拉电阻

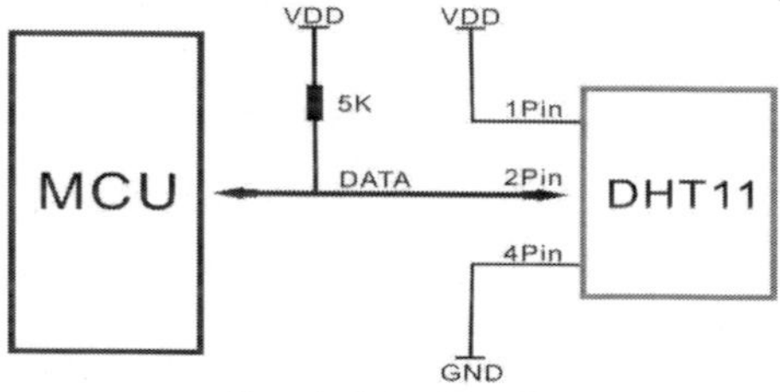

典型应用电路

3、 电源引脚

DHT11的供电电压为 3—5.5V。传感器上电后，要等待 1s 以越过不稳定状态在此期间无需发送任何指令。电源引脚（VDD，GND）之间可增加一个100nF 的电容，用以去耦滤波。

4、串行接口 （单线双向）

DATA 用于微处理器与 DHT11之间的通讯和同步，采用单总线数据格式，一次通讯时间4ms左右，数据分小数部分和整数部分，具体格式在下面说明，当前小数部分用于以后扩展，现读出为零.操作流程如下：

一次完整的数据传输为40bit，高位先出。

数据格式：8bit湿度整数数据+8bit湿度小数数据
+8bi温度整数数据+8bit温度小数数据
+8bit校验和

数据传送正确时校验和数据等于 **“8bit湿度整数数据+8bit湿度小数数据+8bi温度整数数据+8bit温度小数数据”** 所得结果的末8位。

用户MCU发送一次开始信号后，DHT11从低功耗模式转换到高速模式，等待主机开始信号结束后，DHT11发送响应信号，送出40bit的数据，并触发一次信号采集，用户可选择读取部分数据.从模式下，DHT11接收到开始信号触发一次温湿度采集，如果没有接收到主机发送开始信号，DHT11不会主动进行温湿度采集.采集数据后转换到低速模式。

1.通讯过程如图1所示

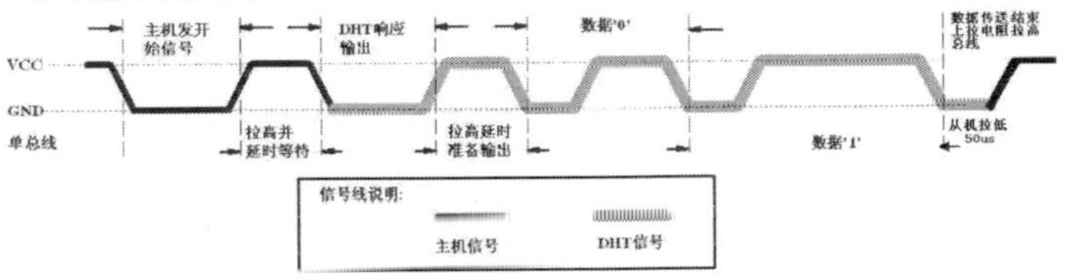

图1

总线空闲状态为高电平，主机把总线拉低等待DHT11响应，主机把总线拉低必须大于18毫秒，保证DHT11能检测到起始信号。DHT11接收到主机的开始信号后，等待主机开始信号结束，然后发送80us低电平响应信号.主机发送开始信号结束后，延时等待20-40us后，读取DHT11的响应信号，主机发送开始信号后，可以切换到输入模式，或者输出高电平均可，总线由上拉电阻拉高。

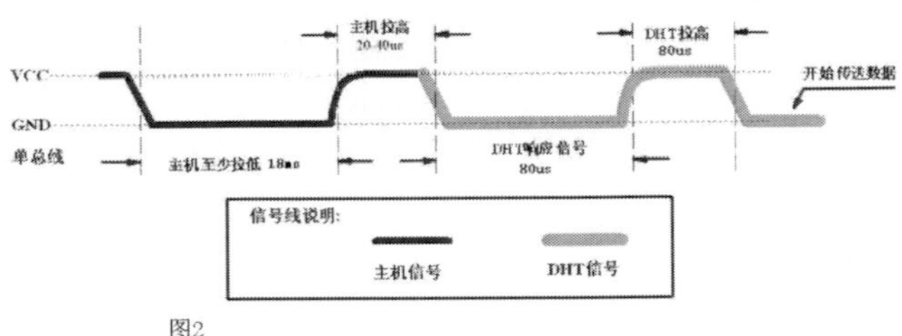

图2

总线为低电平，说明DHT11发送响应信号，DHT11发送响应信号后，再把总线拉高80us，准备发送数据，每一bit数据都以50us低电平时隙开始，高电平的长短定了数据位是0还是1.格式见下面图示.如果读取响应信号为高电平，则DHT11没有响应，请检查线路是否连接正常.当最后一bit数据传送完毕后，DHT11拉低总线50us，随后总线由上拉电阻拉高进入空闲状态。

数字0信号表示方法如图4所示

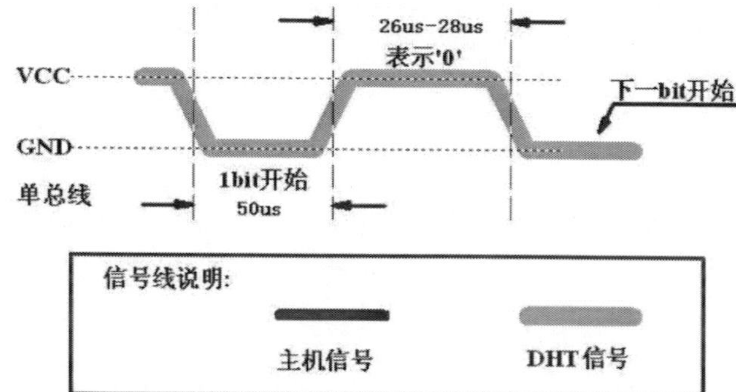

图4

数字1信号表示方法.如图5所示

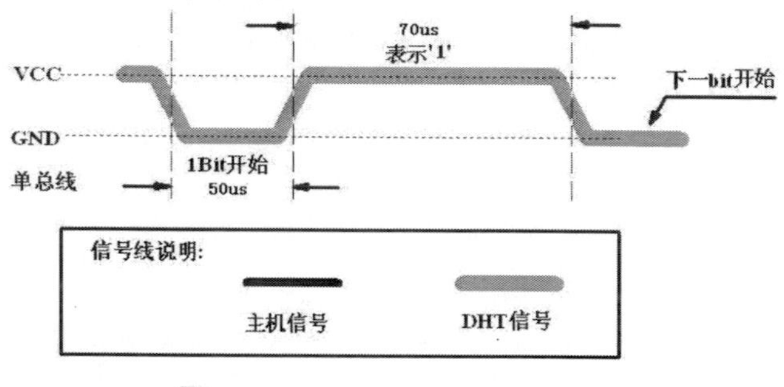

图5

5、 测量分辨率

测量分辨率分别为 8bit（温度）、8bit（湿度）。

6、电气特性

VDD=5V，T＝25℃，除非特殊标注

参数	条件	min	typ	max	单位
供电	DC	3	5	5.5	V
供电电流	测量	0.5		2.5	mA
	平均	0.2		1	mA
	待机	100		150	uA
采样周期	秒	1			次

注:采样周期间隔不得低于1秒钟。

7、应用信息

7.1工作与贮存条件

超出建议的工作范围可能导致高达3%RH的临时性漂移信号。返回正常工作条后，传感器会缓慢地向校准状态恢复。要加速恢复进程/可参阅7.3小节的“恢复处理”。在非正常工作条件下长时间使用会加速产品的老化过程。

7.2暴露在化学物质中

电阻式湿度传感器的感应层会受到化学蒸汽的干扰，化学物质在感应层中的扩散可能导致测量值漂移和灵敏度下降。在一个纯净的环境中，污染物质会缓慢地释放出去。下文所述的恢复处理将加速实现这一过程。高浓度的化学污染会导致传感器感应层的彻底损坏。

7.3恢复处理

置于极限工作条件下或化学蒸汽中的传感器，通过如下处理程序，可使其恢复到校准时的状态。在50-60℃和< 10%RH的湿度条件下保持2 小时（烘干）；随后在20-30℃和>70%RH的湿度条件下保持 5小时以上。

7.4温度影响

气体的相对湿度，在很大程度上依赖于温度。因此在测量湿度时，应尽可能保证湿度传感器在同一温度下工作。如果与释放热量的电子元件共用一个印刷线路板，在安装时应尽可能将DHT11远离电子元件，并安装在热源下方，同时保持外壳的良好通风。为降低热传导，DHT11与印刷电路板其它部分的铜镀层应尽可能最小，并在两者之间留出一道缝隙。

7.5光线

长时间暴露在太阳光下或强烈的紫外线辐射中，会使性能降低。

7.6配线注意事项

DATA信号线材质量会影响通讯距离和通讯质量，推荐使用高质量屏蔽线。

8、封装信息

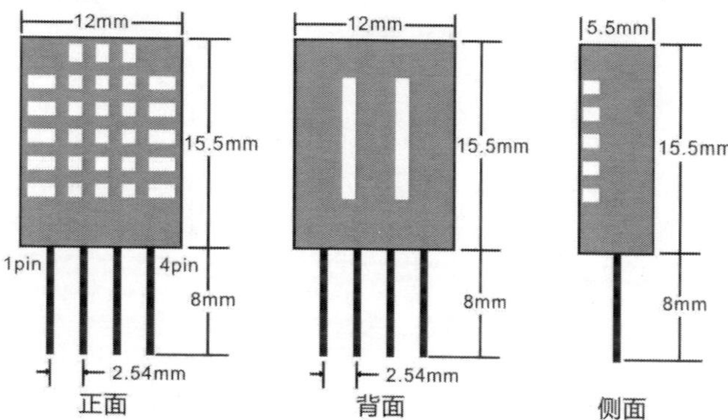

9、 DHT11引脚说明

Pin	名称	注释
1	VDD	供电 3－5.5VDC
2	DATA	串行数据，单总线
3	NC	空脚，请悬空
4	GND	接地，电源负极

10、 焊接信息

手动焊接，在最高260℃的温度条件下接触时间须少于10秒。

11、注意事项

(1)避免结露情况下使用。
(2)长期保存条件：温度10－40℃，湿度60％以下。

資料來源：Adafruit 官網：https://learn.adafruit.com/dht/downloads

DHT21/22 Datasheet

Aosong(Guangzhou) Electronics Co.,Ltd

Tell: +86-020-36380552, +86-020-36042809 Fax: +86-020-36380562
http://www.aosong.com
Email: thomasliu198518@yahoo.com.cn sales@aosong.com
Address: No.56, Renhe Road, Renhe Town, Baiyun District, Guangzhou, China

Digital-output relative humidity & temperature sensor/module

AM2303

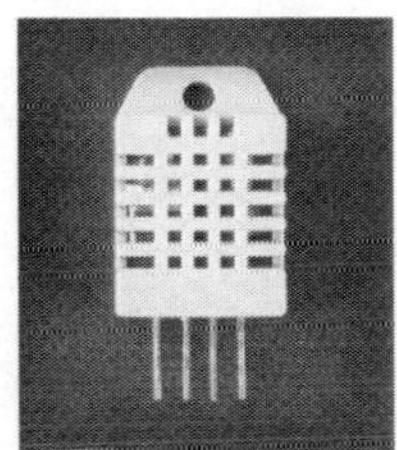

Capacitive-type humidity and temperature module/sensor

1. Feature & Application:

* Full range temperature compensated
* Relative humidity and temperature measurement
* Calibrated digital signal
* Outstanding long-term stability
* Extra components not needed
* Long transmission distance
* Low power consumption
* 4 pins packaged and fully interchangeable

2. Description:

AM2303 output calibrated digital signal. It utilizes exclusive digital-signal-collecting-technique and humidity sensing technology, assuring its reliability and stability.Its sensing elements is connected with 8-bit single-chip computer.

Every sensor of this model is temperature compensated and calibrated in accurate calibration chamber and the calibration-coefficient is saved in type of programme in OTP memory, when the sensor is detecting, it will cite coefficient from memory.
Small size & low consumption & long transmission distance(20m) enable AM2303 to be suited in all kinds of harsh application occasions.

Single-row packaged with four pins, making the connection very convenient.

3. Technical Specification:

Model	AM2303
Power supply	3.3-6V DC
Output signal	digital signal via single-bus
Sensing element	Polymer humidity capacitor & DS18B20 for detecting temperature
Measuring range	humidity 0-100%RH; temperature -40~125Celsius

- 1 -

Tell: +86-020-36380552, +86-020-36042809 Fax: +86-020-36380562
http://www.aosong.com
Email: thomasliu198518@yahoo.com.cn sales@aosong.com
Address: No.56, Renhe Road, Renhe Town, Baiyun District, Guangzhou, China

Accuracy	humidity +-2%RH(Max +-5%RH);	temperature +-0.2Celsius
Resolution or sensitivity	humidity 0.1%RH;	temperature 0.1Celsius
Repeatability	humidity +-1%RH;	temperature +-0.2Celsius
Humidity hysteresis	+-0.3%RH	
Long-term Stability	+-0.5%RH/year	
Sensing period	Average: 2s	
Interchangeability	fully interchangeable	

4. Dimensions: (unit----mm)

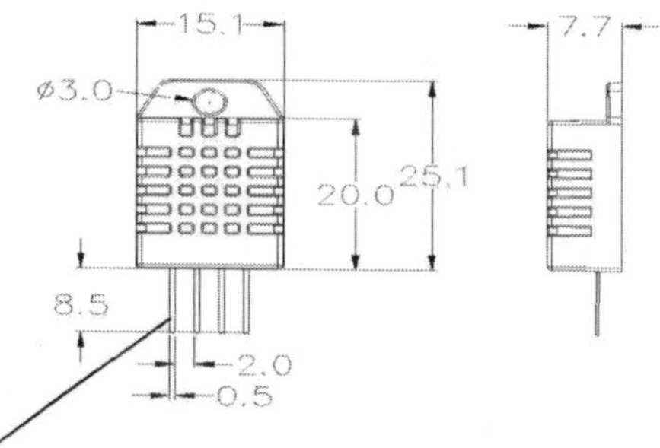

Pin sequence number: 1 2 3 4 (from left to right direction).

Pin	Function
1	VDD---power supply
2	DATA--signal
3	NULL
4	GND

5. Operating specifications:

(1) Power and Pins

Power's voltage should be 3.3-6V DC. When power is supplied to sensor, don't send any instruction to the sensor within one second to pass unstable status. One capacitor valued 100nF can be added between VDD and GND for wave filtering.

(2) Communication and signal

Single-bus data is used for communication between MCU and AM2303, it costs 5mS for single time communication.

Tell: +86-020-36380552, +86-020-36042809 Fax: +86-020-36380562
http://www.aosong.com
Email: thomasliu198518@yahoo.com.cn sales@aosong.com
Address: No.56, Renhe Road, Renhe Town, Baiyun District, Guangzhou, China

Data is comprised of integral and decimal part, the following is the formula for data.

AM2303 send out higher data bit firstly!

DATA=8 bit integral RH data+8 bit decimal RH data+8 bit integral T data+8 bit decimal T data+8 bit check-sum

If the data transmission is right, check-sum should be the last 8 bit of "8 bit integral RH data+8 bit decimal RH data+8 bit integral T data+8 bit decimal T data".

When MCU send start signal, AM2303 change from low-power-consumption-mode to running-mode. When MCU finishs sending the start signal, AM2303 will send response signal of 40-bit data that reflect the relative humidity and temperature information to MCU. Without start signal from MCU, AM2303 will not give response signal to MCU. One start signal for one time's response data that reflect the relative humidity and temperature information from AM2303. AM2303 will change to low-power-consumption-mode when data collecting finish if it don't receive start signal from MCU again.

1) Check bellow picture for overall communication process:

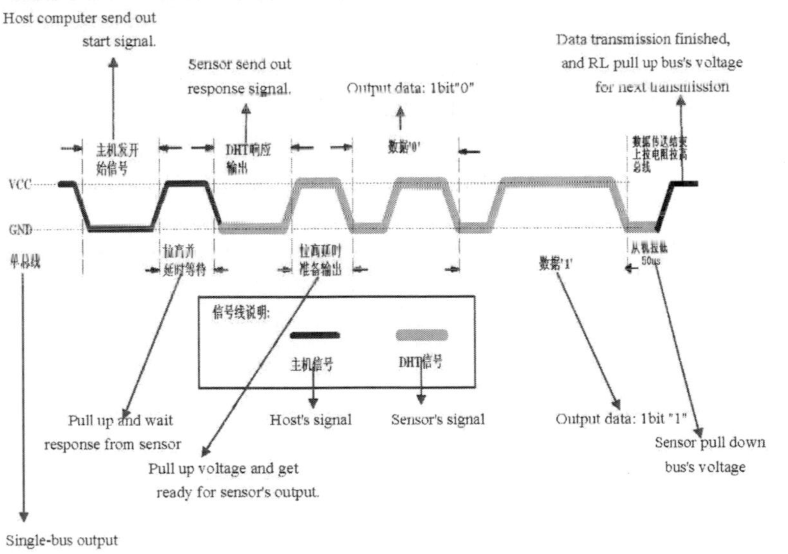

Tell: +86-020-36380552, +86-020-36042809 Fax: +86-020-36380562
http://www.aosong.com
Email: thomasliu198518@yahoo.com.cn sales@aosong.com
Address: No.56, Renhe Road, Renhe Town, Baiyun District, Guangzhou, China

2) Step 1: MCU send out start signal to AM2303

Data-bus's free status is high voltage level. When communication between MCU and AM2303 begin, program of MCU will transform data-bus's voltage level from high to low level and this process must beyond at least 18ms to ensure AM2303 could detect MCU's signal, then MCU will wait 20-40us for AM2303's response.

Check bellow picture for step 1:

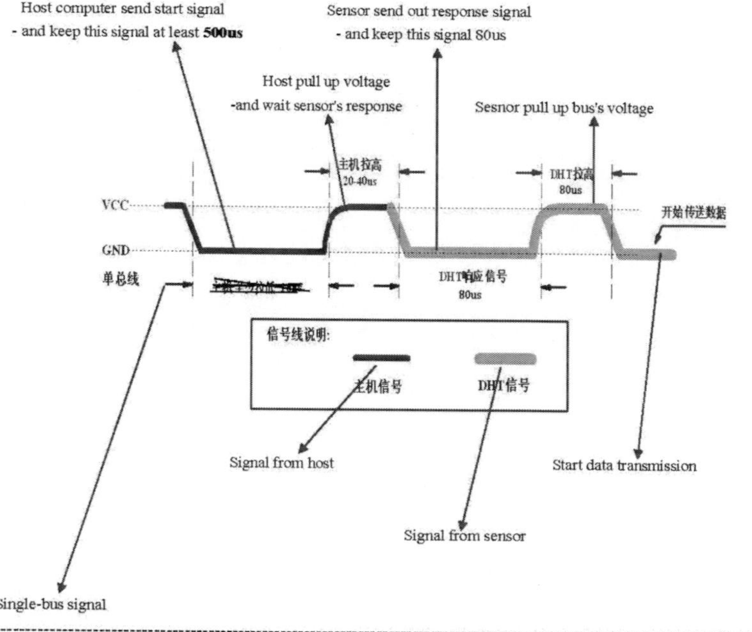

Step 2: AM2303 send response signal to MCU

When AM2303 detect the start signal, AM2303 will send out low-voltage-level signal and this signal last 80us as response signal, then program of AM2303 transform data-bus's voltage level from low to high level and last 80us for AM2303's preparation to send data.

Check bellow picture for step 2:

Tell: +86-020-36380552, +86-020-36042809 Fax: +86-020-36380562
http://www.aosong.com
Email: thomasliu198518@yahoo.com.cn sales@aosong.com
Address: No.56, Renhe Road, Renhe Town, Baiyun District, Guangzhou, China

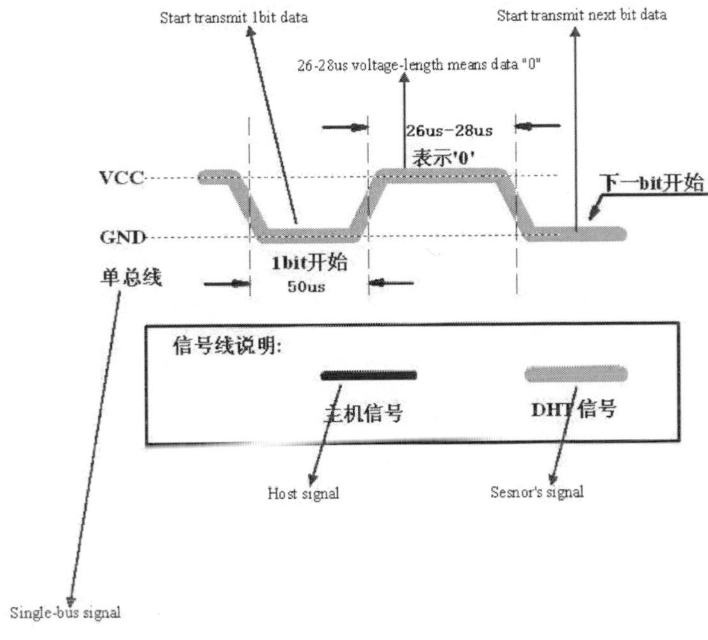

Step 3: AM2303 send data to MCU

When AM2303 is sending data to MCU, every bit's transmission begin with low-voltage-level that last 50us, the following high-voltage-level signal's length decide the bit is "1" or "0".

Check bellow picture for step 3:

Tell: +86-020-36380552, +86-020-36042809 Fax: +86-020-36380562
http://www.aosong.com
Email: thomasliu198518@yahoo.com.cn sales@aosong.com
Address: No.56, Renhe Road, Renhe Town, Baiyun District, Guangzhou, China

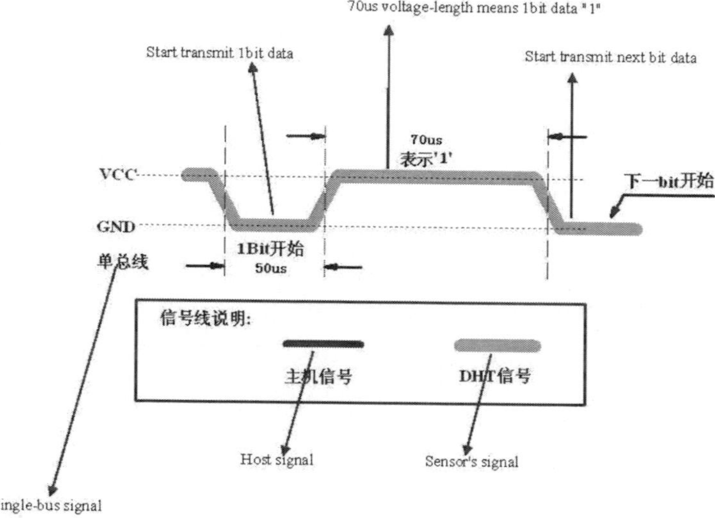

If signal from AM2303 is always high-voltage-level, it means AM2303 is not working properly, please check the electrical connection status.

6. Electrical Characteristics:

Item	Condition	Min	Typical	Max	Unit
Power supply	DC	3.3	5	5.5	V
Current supply	Measuring	1.3	1.5	2.1	mA
	Average	0.5	0.8	1.1	mA
Collecting period	Second	1.7		2	Second

*Collecting period should be : >1.7 second.

7. Attentions of application:

(1) Operating and storage conditions

We don't recommend the applying RH-range beyond the range stated in this specification. The DHT11 sensor

Tell: +86-020-36380552, +86-020-36042809 Fax: +86-020-36380562
http://www.aosong.com
Email: thomasliu198518@yahoo.com.cn sales@aosong.com
Address: No.56, Renhe Road, Renhe Town, Baiyun District, Guangzhou, China

can recover after working in non-normal operating condition to calibrated status, but will accelerate sensors' aging.

(2) Attentions to chemical materials

Vapor from chemical materials may interfere AM2303's sensitive-elements and debase AM2303's sensitivity.

(3) Disposal when (1) & (2) happens

Step one: Keep the AM2303 sensor at condition of Temperature 50~60Celsius, humidity <10%RH for 2 hours;

Step two: After step one, keep the AM2303 sensor at condition of Temperature 20~30Celsius, humidity >70%RH for 5 hours.

(4) Attention to temperature's affection

Relative humidity strongly depend on temperature, that is why we use temperature compensation technology to ensure accurate measurement of RH. But it's still be much better to keep the sensor at same temperature when sensing.

AM2303 should be mounted at the place as far as possible from parts that may cause change to temperature.

(5) Attentions to light

Long time exposure to strong light and ultraviolet may debase AM2303's performance.

(6) Attentions to connection wires

The connection wires' quality will effect communication's quality and distance, high quality shielding-wire is recommended.

(7) Other attentions

* Welding temperature should be bellow 260Celsius.

* Avoid using the sensor under dew condition.

* Don't use this product in safety or emergency stop devices or any other occasion that failure of AM2303 may cause personal injury.

SHT21 Datasheet

SENSIRION
THE SENSOR COMPANY

Datasheet SHT21

Humidity and Temperature Sensor IC

- Fully calibrated
- Digital output, I²C interface
- Low power consumption
- Excellent long-term stability
- DFN type package – reflow solderable

Product Summary

The SHT21 humidity and temperature sensor of Sensirion has become an industry standard in terms of form factor and intelligence: Embedded in a reflow solderable Dual Flat No leads (DFN) package of 3 x 3mm foot print and 1.1mm height it provides calibrated, linearized sensor signals in digital, I²C format.

The SHT2x sensors contain a capacitive type humidity sensor, a band gap temperature sensor and specialized analog and digital integrated circuit – all on a single CMOSens® chip. This yields in an unmatched sensor performance in terms of accuracy and stability as well as minimal power consumption.

Every sensor is individually calibrated and tested. Lot identification is printed on the sensor and an electronic identification code is stored on the chip – which can be read out by command. Furthermore, the resolution of SHT2x can be changed by command (8/12bit up to 12/14bit for RH/T) and a checksum helps to improve communication reliability.

With this set of features and the proven reliability and long-term stability, the SHT2x sensors offer an outstanding performance-to-price ratio. For testing SHT2x two evaluation kits EK-H4 and EK-H5 are available.

Dimensions

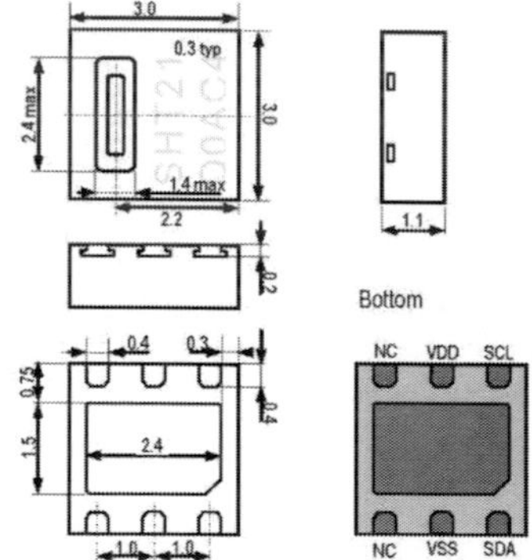

Figure 1: Drawing of SHT21 sensor package, dimensions are given in mm (1mm = 0.039inch), tolerances are ±0.1mm. The die pad (center pad) is internally connected to VSS. The NC pads must be left floating. VSS = GND, SDA = DATA. Numbering of E/O pads starts at lower right corner (indicated by notch in die pad) and goes clockwise (compare Table 2).

Sensor Chip

SHT21 features a generation 4C CMOSens® chip. Besides the capacitive relative humidity sensor and the band gap temperature sensor, the chip contains an amplifier, A/D converter, OTP memory and a digital processing unit.

Material Contents

While the sensor itself is made of Silicon the sensors' housing consists of a plated Cu lead-frame and green epoxy-based mold compound. The device is fully RoHS and WEEE compliant, e.g. free of Pb, Cd and Hg.

Additional Information and Evaluation Kits

Additional information such as Application Notes is available from the web page www.sensirion.com/sht21. For more information please contact Sensirion via info@sensirion.com.

For SHT2x two Evaluation Kits are available: EK-H4, a four-channel device with Viewer Software, that also serves for data-logging, and a simple EK-H5 directly connecting one sensor via USB port to a computer.

Sensor Performance

Relative Humidity

Parameter	Condition	Value	Units
Resolution [1]	12 bit	0.04	%RH
	8 bit	0.7	%RH
Accuracy tolerance [2]	typ	±2	%RH
	max	see Figure 2	%RH
Repeatability		±0.1	%RH
Hysteresis		±1	%RH
Nonlinearity		<0.1	%RH
Response time [3]	τ 63%	8	s
Operating Range	extended [4]	0 to 100	%RH
Long Term Drift [5]	Typ.	< 0.25	%RH/yr

Temperature

Parameter	Condition	Value	Units
Resolution [1]	14 bit	0.01	°C
	12 bit	0.04	°C
Accuracy tolerance [2]	typ	±0.3	°C
	max	see Figure 3	
Repeatability		±0.1	°C
Operating Range	extended [4]	-40 to 125	°C
Response Time [7]	τ 63%	5 to 30	s
Long Term Drift [8]	Typ.	< 0.02	°C/yr

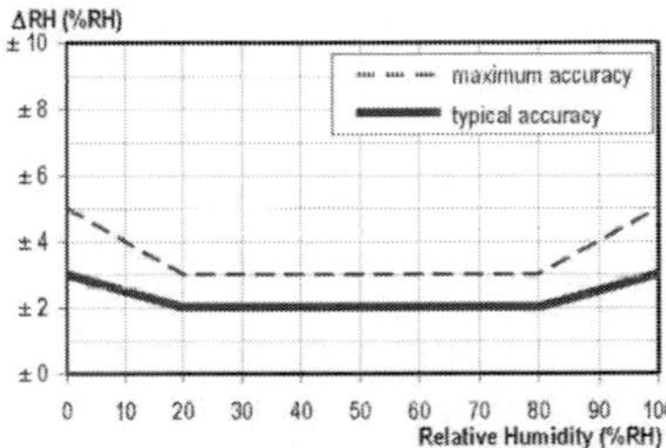

Figure 2 Typical and maximal tolerance at 25°C for relative humidity. For extensive information see Users Guide, Sect. 1.2.

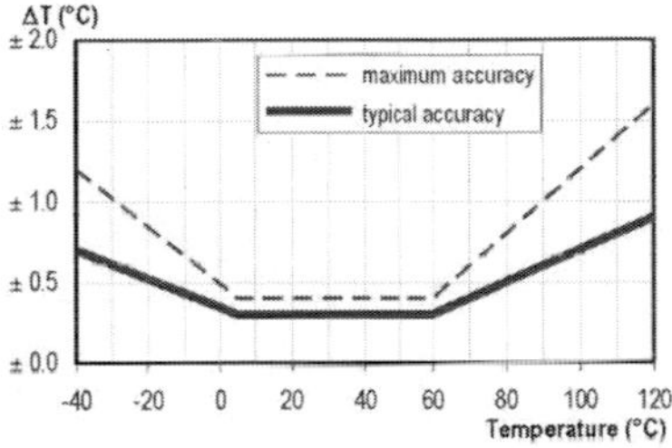

Figure 3 Typical and maximal tolerance for temperature sensor in °C.

Electrical Specification

Parameter	Condition	min	typ	max	Units
Supply Voltage, VDD		2.1	3.0	3.6	V
Supply Current, IDD [6]	sleep mode		0.15	0.4	μA
	measuring	200	300	330	μA
Power Dissipation [6]	sleep mode		0.5	1.2	μW
	measuring	0.6	0.9	1.0	mW
	average 8bit		3.2		μW
Heater	VDD = 3.0 V	5.5mW, ΔT = + 0.5-1.5°C			
Communication	digital 2-wire interface, I²C protocol				

Table 1 Electrical specification. For absolute maximum values see Section 4.1 of Users Guide.

Packaging Information

Sensor Type	Packaging	Quantity	Order Number
SHT21	Tape & Reel	400	1-100707-01
	Tape & Reel	1500	1-100645-01
	Tape & Reel	5000	1-100694-01

This datasheet is subject to change and may be amended without prior notice

[1] Default measurement resolution is 14bit (temperature) / 12bit (humidity). It can be reduced to 12/8bit, 11/11bit or 13/10bit by command to user register.

[2] Accuracies are tested at Outgoing Quality Control at 25°C and 3.0V. Values exclude hysteresis and long term drift and are applicable to non-condensing environments only.

[3] Time for achieving 63% of a step function, valid at 25°C and 1m/s airflow.

[4] Normal operating range: 0-80%RH, beyond this limit sensor may read a reversible offset with slow kinetics (+3%RH after 60h at humidity >80%RH). For more details please see Section 1.1 of the Users Guide.

[5] Typical value for operation in normal RH/T operating range. Max. value is < 0.5 %RH/y. Value may be higher in environments with vaporized solvents, out-gassing tapes, adhesives, packaging materials, etc. For more details please refer to Handling Instructions.

[6] Min and max values of Supply Current and Power Dissipation are based on fixed VDD = 3.0V and T<60°C. The average value is based on one 8bit measurement per second.

[7] Response time depends on heat conductivity of sensor substrate.

[8] Max. value is < 0.04°C/y.

Users Guide SHT21

1 Extended Specification

For details on how Sensirion is specifying and testing accuracy performance please consult Application Note "Statement on Sensor Specification".

1.1 Operating Range

The sensor works stable within recommended Normal Range – see Figure 4. Long term exposure to conditions outside Normal Range, especially at humidity >80%RH, may temporarily offset the RH signal (+3%RH after 60h). After return into the Normal Range it will slowly return towards calibration state by itself. Prolonged exposure to extreme conditions may accelerate ageing.

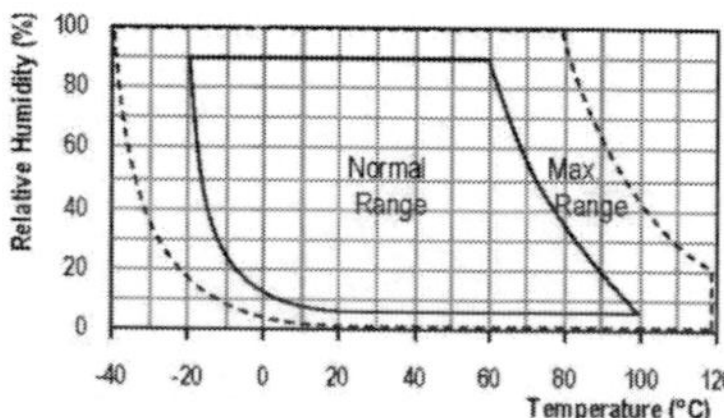

Figure 4 Operating Conditions

1.2 RH accuracy at various temperatures

Typical RH accuracy at 25°C is defined in Figure 2. For other temperatures, typical accuracy has been evaluated to be as displayed in Figure 5.

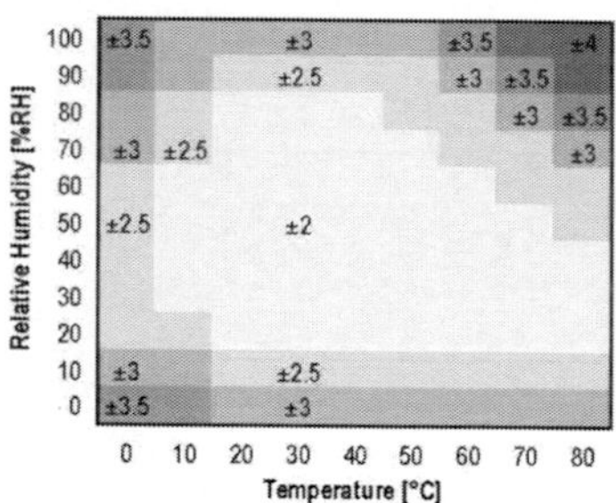

Figure 5 Typical accuracy of relative humidity measurements given in %RH for temperatures 0 – 80°C.

1.3 Electrical Specification

Current consumption as given in Table 1 is dependent on temperature and supply voltage VDD. For estimations on energy consumption of the sensor Figures 6 and 7 may be consulted. Please note that values given in these Figures are of typical nature and the variance is considerable.

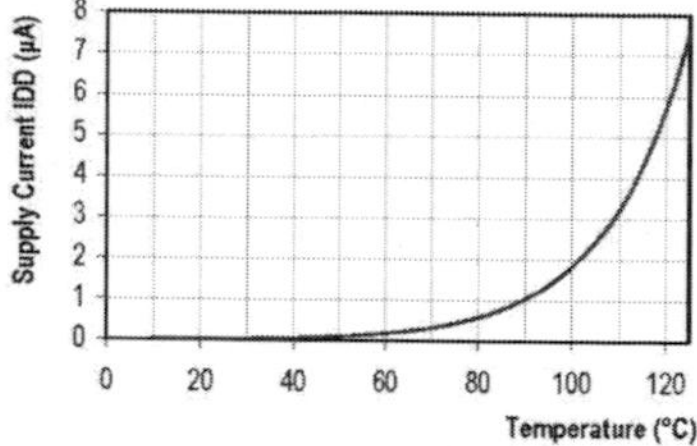

Figure 6 Typical dependency of supply current (sleep mode) versus temperature at VDD = 3.0V. Please note that the variance of these data can be above ±25% of displayed value.

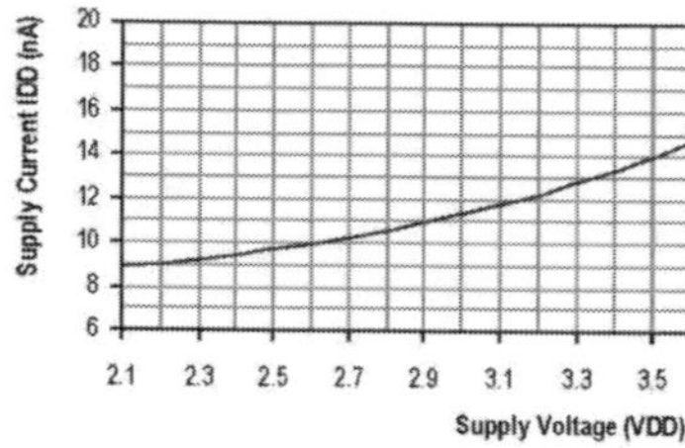

Figure 7 Typical dependency of supply current (sleep mode) versus supply voltage at 25°C. Please note that deviations may be up to ±50% of displayed value. Values at 60°C scale with a factor of about 15 (compare Table 1).

2 Application Information

2.1 Soldering Instructions

The *DFN's die pad* (centre pad) and *perimeter I/O pads* are fabricated from a planar copper lead-frame by over-molding leaving the die pad and I/O pads exposed for mechanical and electrical connection. Both the I/O pads and die pad should be soldered to the PCB. In order to prevent oxidation and optimize soldering, the bottom side of the sensor pads is plated with Ni/Pd/Au.

On the PCB the *I/O lands*[9] should be 0.2mm longer than the package I/O pads. Inward corners may be rounded to match the I/O pad shape. The I/O land width should match the DFN-package I/O-pads width 1:1 and the land for the die pad should match 1:1 with the DFN package – see Figure 8.

The *solder mask*[10] *design* for the land pattern preferably is of type Non-Solder Mask Defined (NSMD) with solder mask openings larger than metal pads. For NSMD pads, the solder mask opening should be about 120µm to 150µm larger than the pad size, providing a 60µm to 75µm design clearance between the copper pad and solder mask. Rounded portions of package pads should have a matching rounded solder mask-opening shape to minimize the risk of solder bridging. For the actual pad dimensions, each pad on the PCB should have its own solder mask opening with a web of solder mask between adjacent pads.

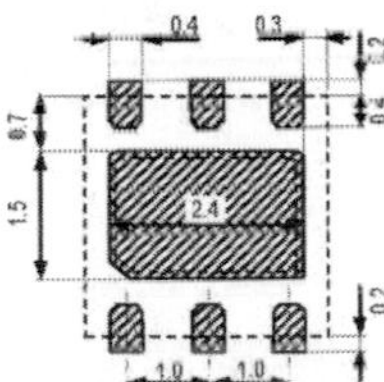

Figure 8 Recommended metal land pattern for SHT2x. Values in mm. Die pad (centre pad) may be left floating or be connected to ground, NC pads shall be left floating. The outer dotted line represents the outer dimension of the DFN package.

For *solder paste printing* a laser-cut, stainless steel stencil with electro-polished trapezoidal walls and with 0.125mm stencil thickness is recommended. For the I/O pads the stencil apertures should be 0.1mm longer than PCB pads and positioned with 0.1mm offset away from the centre of the package. The die pad aperture should cover about 70 – 90% of the pad area – say up to 1.4mm x 2.3mm centered on the thermal land area. It can also be split in two openings.

Due to the low mounted height of the DFN, "no clean" type 3 solder paste[11] is recommended as well as Nitrogen purge during reflow.

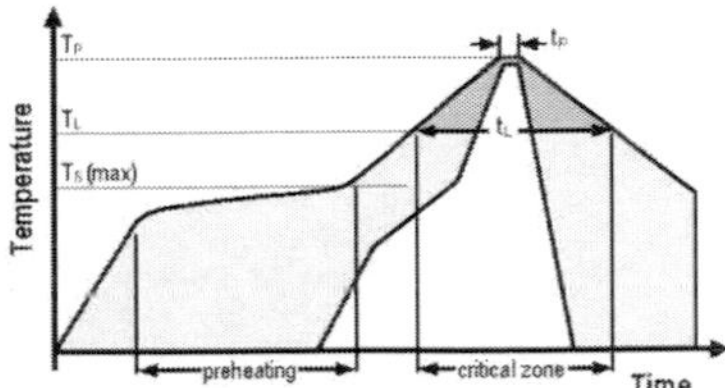

Figure 9 Soldering profile according to JEDEC standard. T_P <= 260°C and t_P < 30sec for Pb-free assembly. T_L < 220°C and t_L < 150sec. Ramp-up/down speeds shall be < 5°C/sec.

It is important to note that the diced edge or side faces of the I/O pads may oxidise over time, therefore a solder fillet may or may not form. Hence there is no guarantee for solder joint fillet heights of any kind.

For soldering SHT2x, standard *reflow soldering* ovens may be used. The sensor is qualified to withstand soldering profile according to IPC/JEDEC J-STD-020 with peak temperatures at 260°C during up to 30sec for Pb-free assembly in IR/Convection reflow ovens (see Figure 9).

For manual soldering contact time must be limited to 5 seconds at up to 350°C.

Immediately after the exposure to high temperatures the sensor may temporarily read a negative humidity offset (typ. -1 to -2 %RH after reflow soldering). This offset slowly disappears again by itself when the sensor is exposed to ambient conditions (typ. within 1-3 days). If RH testing is performed immediately after reflow soldering, this offset should be considered when defining the test limits.

In no case, neither after manual nor reflow soldering, a board wash shall be applied. Therefore, and as mentioned above, it is strongly recommended to use "no-clean" solder paste. In case of applications with exposure of the sensor to corrosive gases or condensed water (i.e. environments with high relative humidity) the soldering pads shall be sealed (e.g. conformal coating) to prevent loose contacts or short cuts.

2.2 Storage Conditions and Handling Instructions

Moisture Sensitivity Level (MSL) is 1, according to IPC/JEDEC J-STD-020. At the same time, it is

[9] The land pattern is understood to be the metal layer on the PCB, onto which the DFN pads are soldered to.

[10] The solder mask is understood to be the insulating layer on top of the PCB covering the connecting lines.

[11] Solder types are related to the solder particle size in the paste: Type 3 covers the size range of 25 – 45 µm (powder type 42).

recommended to further process the sensors within 1 year after date of delivery.

It is of great importance to understand that a humidity sensor is not a normal electronic component and needs to be handled with care. Chemical vapors at high concentration in combination with long exposure times may offset the sensor reading.

For this reason it is recommended to store the sensors in original packaging including the sealed ESD bag at following conditions: Temperature shall be in the range of 10°C – 50°C and humidity at 20 – 60%RH (sensors that are not stored in ESD bags). For sensors that have been removed from the original packaging we recommend to store them in ESD bags made of metal-in PE-HD[12].

In manufacturing and transport the sensors shall be prevented of high concentration of chemical solvents and long exposure times. Out-gassing of glues, adhesive tapes and stickers or out-gassing packaging material such as bubble foils, foams, etc. shall be avoided. Manufacturing area shall be well ventilated.

For more detailed information please consult the document *"Handling Instructions"* or contact Sensirion.

2.3 Temperature Effects

Relative humidity reading strongly depends on temperature. Therefore, it is essential to keep humidity sensors at the same temperature as the air of which the relative humidity is to be measured. In case of testing or qualification the reference sensor and test sensor must show equal temperature to allow for comparing humidity readings.

If the sensor shares a PCB with electronic components that produce heat it should be mounted in a way that prevents heat transfer or keeps it as low as possible. Measures to reduce heat transfer can be ventilation, reduction of copper layers between the sensor and the rest of the PCB or milling a slit into the PCB around the sensor – see Figure 10.

Furthermore, there are self-heating effects in case the measurement frequency is too high. To keep self-heating below 0.1°C, SHT2x should not be active for more than 10% of the time – e.g. maximum two measurements per second at 12bit accuracy shall be made.

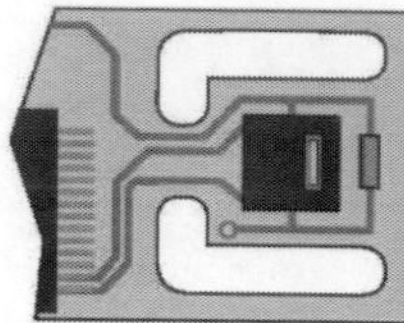

Figure 10 Top view of example of mounted SHT2x with slits milled into PCB to minimize heat transfer.

2.4 Light

The SHT2x is not light sensitive. Prolonged direct exposure to sunshine or strong UV radiation may age the sensor.

2.5 Materials Used for Sealing / Mounting

Many materials absorb humidity and will act as a buffer increasing response times and hysteresis. Materials in the vicinity of the sensor must therefore be carefully chosen. Recommended materials are: Any metals, LCP, POM (Delrin), PTFE (Teflon), PEEK, PP, PB, PPS, PSU, PVDF, PVF.

For sealing and gluing (use sparingly): Use high filled epoxy for electronic packaging (e.g. glob top, underfill), and Silicone. Out-gassing of these materials may also contaminate the sensor (see Section 2.2). Therefore try to add the sensor as a last manufacturing step to the assembly, store the assembly well ventilated after manufacturing or bake at >50°C for 24h to outgas contaminants before packing.

2.6 Wiring Considerations and Signal Integrity

Carrying the SCL and SDA signal parallel and in close proximity (e.g. in wires) for more than 10cm may result in cross talk and loss of communication. This may be resolved by routing VDD and/or VSS between the two SDA signals and/or using shielded cables. Furthermore, slowing down SCL frequency will possibly improve signal integrity. Power supply pins (VDD, VSS) must be decoupled with a 100nF capacitor – see next Section.

[12] For example, 3M antistatic bag, product "1910" with zipper.

3 Interface Specifications

Pin	Name	Comment
1	SDA	Serial Data, bidirectional
2	VSS	Ground
5	VDD	Supply Voltage
6	SCL	Serial Clock, bidirectional
3,4	NC	Not Connected

Table 2 SHT2x pin assignment, NC remain floating (top view)

3.1 Power Pins (VDD, VSS)

The supply voltage of SHT2x must be in the range of 2.1 – 3.6V, recommended supply voltage is 3.0V. Power supply pins Supply Voltage (VDD) and Ground (VSS) must be decoupled with a 100nF capacitor, that shall be placed as close to the sensor as possible – see Figure 11.

3.2 Serial clock (SCL)

SCL is used to synchronize the communication between microcontroller (MCU) and the sensor. Since the interface consists of fully static logic there is no minimum SCL frequency.

3.3 Serial SDA (SDA)

The SDA pin is used to transfer data in and out of the sensor. For sending a command to the sensor, SDA is valid on the rising edge of SCL and must remain stable while SCL is high. After the falling edge of SCL the SDA value may be changed. For safe communication SDA shall be valid t_{SU} and t_{HD} before the rising and after the falling edge of SCL, respectively – see Figure 12. For reading data from the sensor, SDA is valid t_{VD} after SCL has gone low and remains valid until the next falling edge of SCL.

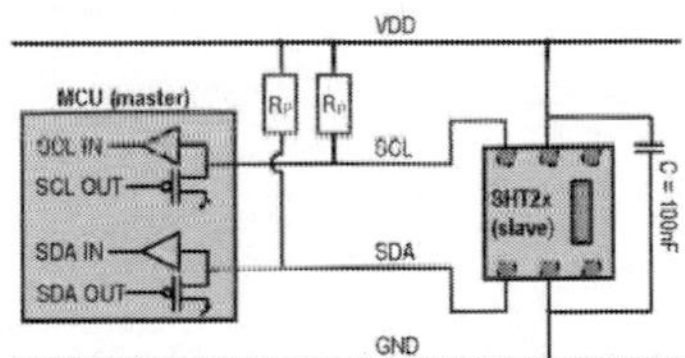

Figure 11 Typical application circuit, including pull-up resistors R_P and decoupling of VDD and VSS by a capacitor.

To avoid signal contention the micro-controller unit (MCU) must only drive SDA and SCL low. External pull-up resistors (e.g. 10kΩ), are required to pull the signal high. For the choice of resistor size please take bus capacity requirements into account (compare Table 5). It should be noted that pull-up resistors may be included in I/O circuits of MCUs. See Table 4 and Table 5 for detailed I/O characteristic of the sensor.

4 Electrical Characteristics

4.1 Absolute Maximum Ratings

The electrical characteristics of SHT2x are defined in Table 1. The absolute maximum ratings as given in Table 3 are stress ratings only and give additional information. Functional operation of the device at these conditions is not implied. Exposure to absolute maximum rating conditions for extended periods may affect the sensor reliability (e.g. hot carrier degradation, oxide breakdown).

Parameter	min	max	Units
VDD to VSS	-0.3	5	V
Digital I/O Pins (SDA, SCL) to VSS	-0.3	VDD + 0.3	V
Input Current on any Pin	-100	100	mA

Table 3 Electrical absolute maximum ratings

ESD immunity is qualified according to JEDEC JESD22-A114 method (Human Body Model at ±4kV), JEDEC JESD22-A115 method (Machine Model ±200V) and ESDA ESD-STM5.3.1-1999 and AEC-Q100-011 (Charged Device Model, 750V corner pins, 500V other pins). Latch-up immunity is provided at a force current of ±100mA with T_{amb} = 125°C according to JEDEC JESD78. For exposure beyond named limits the sensor needs additional protection circuit.

4.2 Input / Output Characteristics

The electrical characteristics such as power consumption, low and high level input and output voltages depend on the supply voltage. For proper communication with the sensor it is essential to make sure that signal design is strictly within the limits given in Table 4 & 5 and Figure 12.

Parameter	Conditions	min	typ	max	Units
Output Low Voltage, VOL	VDD = 3.0 V, -4 mA < IOL < 0mA	0	-	0.4	V
Output Sink Current, IOL		-	-	-4	mA
Input Low Voltage, VIL		0	-	30% VDD	V
Input High Voltage, VIH		70% VDD	-	VDD	V
Input Current	VDD = 3.6 V, VIN = 0 V to 3.6 V	-	-	±1	uA

Table 4 DC characteristics of digital input/output pads. VDD = 2.1V to 3.6V, T = -40°C to 125°C, unless otherwise noted.

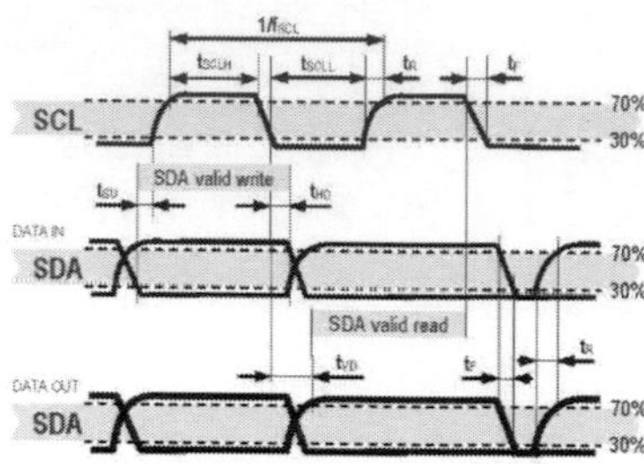

Figure 12 Timing Diagram for Digital Input/Output Pads, abbreviations are explained in Table 5. SDA directions are seen from the sensor. Bold SDA line is controlled by the sensor, plain SDA line is controlled by the micro-controller. Note that SDA valid read time is triggered by falling edge of anterior toggle.

Parameter	min	typ	max	Units
SCL frequency, f_{SCL}	0	-	0.4	MHz
SCL High Time, t_{SCLH}	0.6	-	-	µs
SCL Low Time, t_{SCLL}	1.3	-	-	µs
SDA Set-Up Time, t_{SU}	100	-	-	ns
SDA Hold Time, t_{HD}	0	-	900	ns
SDA Valid Time, t_{VD}	0	-	400	ns
SCL/SDA Fall Time, t_F	0	-	100	ns
SCL/SDA Rise Time, t_R	0	-	300	ns
Capacitive Load on Bus Line, C_B	0	-	400	pF

Table 5 Timing specifications of digital input/output pads for I²C fast mode. Entities are displayed in Figure 12. VDD = 2.1V to 3.6V, T = -40°C to 125°C, unless otherwise noted. For further information regarding timing, please refer to http://www.standardics.nxp.com/support/i2c/.

5 Communication with Sensor

SHT21 communicates with I²C protocol. For information on I²C beyond the information in the following Sections please refer to the following website:

http://www.standardics.nxp.com/support/i2c/.

Please note that all sensors are set to the same I²C address, as defined in Section 5.3.

Furthermore, please note, that Sensirion provides an exemplary sample code on its home page – compare www.sensirion.com/sht21.

Please note that in case VDD is set to 0 V (GND), e.g. in case of a power off of the SHT2x, the SCL and SDA pads are also pulled to GND. Consequently, the I2C bus is blocked while VDD of the SHT2x is set to 0 V.

5.1 Start Up Sensor

As a first step, the sensor is powered up to the chosen supply voltage VDD (between 2.1V and 3.6V). After power-up, the sensor needs at most 15ms, while SCL is high, for reaching idle state, i.e. to be ready accepting commands from the master (MCU). Current consumption during start up is 350µA maximum. Whenever the sensor is powered up, but not performing a measurement or communicating, it is automatically in idle state (sleep mode).

5.2 Start / Stop Sequence

Each transmission sequence begins with Start condition (S) and ends with Stop condition (P) as displayed in Figure 13 and Figure 14.

Figure 13 Transmission Start condition (S) - a high to low transition on the SDA line while SCL is high. The Start condition is a unique state on the bus created by the master, indicating to the slaves the beginning of a transmission sequence (bus is considered busy after a Start).

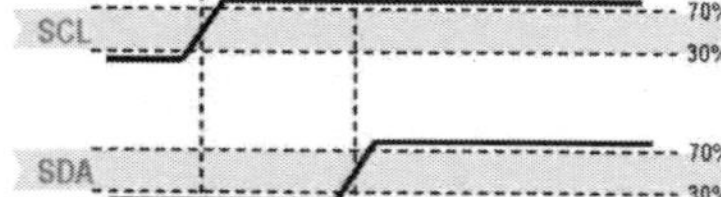

Figure 14 Transmission Stop condition (P) - a low to high transition on the SDA line while SCL is high. The Stop condition is a unique state on the bus created by the master, indicating to the slaves the end of a transmission sequence (bus is considered free after a Stop).

5.3 Sending a Command

After sending the Start condition, the subsequent I²C header consists of the 7-bit I²C device address '1000'000' and an SDA direction bit (Read R: '1', Write W: '0'). The sensor indicates the proper reception of a byte by pulling the SDA pin low (ACK bit) after the falling edge of the 8th SCL clock. After the issue of a measurement command ('1110'0011' for temperature, '1110'0101' for relative humidity'), the MCU must wait for the measurement to complete. The basic commands are summarized in Table 6.

Command	Comment	Code
Trigger T measurement	hold master	1110'0011
Trigger RH measurement	hold master	1110'0101
Trigger T measurement	no hold master	1111'0011
Trigger RH measurement	no hold master	1111'0101
Write user register		1110'0110
Read user register		1110'0111
Soft reset		1111'1110

Table 6 Basic command set, RH stands for relative humidity, and T stands for temperature

Hold master or *no hold master* modes are explained in next Section.

5.4 Hold / No Hold Master Mode

There are two different operation modes to communicate with the sensor: *Hold Master* mode or *No Hold Master* mode. In the first case the SCL line is blocked (controlled by sensor) during measurement process while in the latter case the SCL line remains open for other communication while the sensor is processing the measurement. No hold master mode allows for processing other I²C communication tasks on a bus while the sensor is measuring. A communication sequence of the two modes is displayed in Figure 15 and Figure 16, respectively.

In the *hold master mode*, the SHT2x pulls down the SCL line while measuring to force the master into a wait state. By releasing the SCL line the sensor indicates that internal processing is terminated and that transmission may be continued.

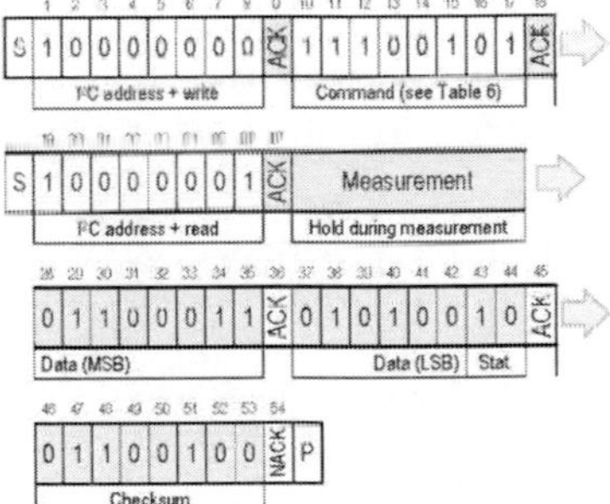

Figure 15 *Hold master* communication sequence – grey blocks are controlled by SHT2x. Bit 45 may be changed to NACK followed by Stop condition (P) to omit checksum transmission.

In *no hold master* mode, the MCU has to poll for the termination of the internal processing of the sensor. This is done by sending a Start condition followed by the I²C header (1000'0001) as shown in Figure 16. If the internal processing is finished, the sensor acknowledges the poll of the MCU and data can be read by the MCU. If the measurement processing is not finished the sensor answers no ACK bit and the Start condition must be issued once more.

When using the *no hold master* mode it is recommended to include a wait period of 20 µs after the reception of the sensor's ACK bit (bit 18 in Figure 16) and before the Stop condition.

For both modes, since the maximum resolution of a measurement is 14 bit, the two last least significant bits (LSBs, bits 43 and 44) are used for transmitting status information. Bit 1 of the two LSBs indicates the measurement type ('0': temperature, '1' humidity). Bit 0 is currently not assigned.

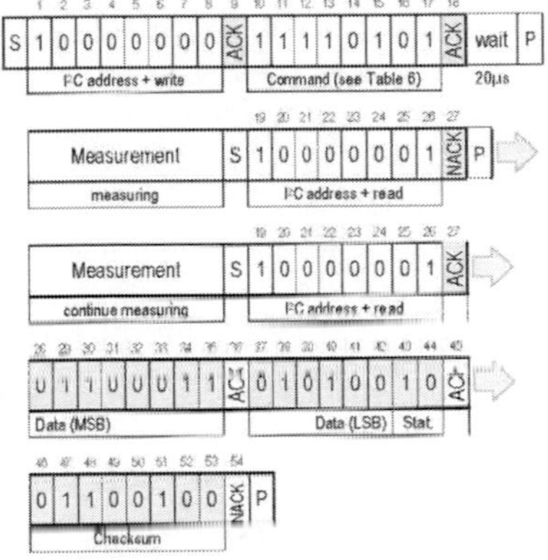

Figure 16 *No Hold master* communication sequence – grey blocks are controlled by SHT2x. If measurement is not completed upon 'read' command, sensor does not provide ACK on bit 27 (more of these iterations are possible). If bit 45 is changed to NACK followed by Stop condition (P) checksum transmission is omitted.

In the examples given in Figure 15 and Figure 16 the sensor output is S_{RH} = '0110'0011'0101'0000'. For the calculation of physical values Status Bits must be set to '0' – see Chapter 6.

The maximum duration for measurements depends on the type of measurement and resolution chosen – values are displayed in Table 7. Maximum values shall be chosen for the communication planning of the MCU.

Resolution	RH typ	RH max	T typ	T max	Units
14 bit			66	85	ms
13 bit			33	43	ms
12 Bit	22	29	17	22	ms
11 bit	12	15	9	11	ms
10 bit	7	9			ms
8 bit	3	4			ms

Table 7 Measurement times for RH and T measurements at different resolutions. Typical values are recommended for calculating energy consumption while maximum values shall be applied for calculating waiting times in communication.

Please note: I²C communication allows for repeated Start conditions (S) without closing prior sequence with Stop condition (P) – compare Figures 15, 16 and 18. Still, any sequence with adjacent Start condition may alternatively be closed with a Stop condition.

5.5 Soft Reset

This command (see Table 6) is used for rebooting the sensor system without switching the power off and on again. Upon reception of this command, the sensor system reinitializes and starts operation according to the default settings – with the exception of the heater bit in the user register (see Sect. 5.6). The soft reset takes less than 15ms.

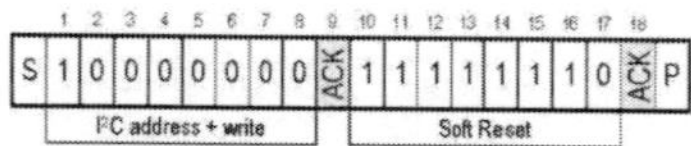

Figure 17 *Soft Reset* – grey blocks are controlled by SHT2x.

5.6 User Register

The content of User Register is described in Table 8. Please note that reserved bits must not be changed and default values of respective reserved bits may change over time without prior notice. Therefore, for any writing to the User Register, default values of reserved bits must be read first. Thereafter, the full User Register string is composed of respective default values of reserved bits and the remainder of accessible bits optionally with default or non-default values.

The *end of battery* alert is activated when the battery power falls below 2.25V.

The *heater* is intended to be used for functionality diagnosis – relative humidity drops upon rising temperature. The heater consumes about 5.5mW and provides a temperature increase of about 0.5 – 1.5°C.

OTP Reload is a safety feature and loads the entire OTP settings to the register, with the exception of the heater bit, before every measurement. This feature is disabled per default and is not recommended for use. Please use Soft Reset instead – it contains OTP Reload.

Bit	# Bits	Description / Coding	Default
7, 0	2	Measurement resolution RH / T '00' 12 bit / 14 bit '01' 8 bit / 12 bit '10' 10 bit / 13 bit '11' 11 bit / 11 bit	'00'
6	1	Status: End of battery[15] '0': VDD > 2.25V '1': VDD < 2.25V	'0'
3, 4, 5	3	Reserved	
2	1	Enable on-chip heater	'0'
1	1	Disable OTP Reload	'1'

Table 8 User Register. Cut-off value for End of Battery signal may vary by ±0.1V. Reserved bits must not be changed. "OTP reload" = '0' loads default settings after each time a measurement command is issued.

An example for I²C communication reading and writing the User Register is given in Figure 18.

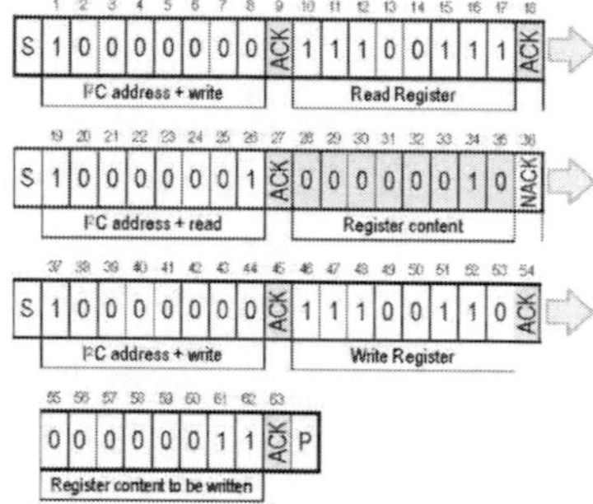

0 0 0 0 0 0 1 1 ACK P

Register content to be written

Figure 18 *Read and write register sequence* – grey blocks are controlled by SHT2x. In this example, the resolution is set to 8bit / 12bit.

5.7 CRC Checksum

SHT21 provides a CRC-8 checksum for error detection. The polynomial used is $x^8 + x^5 + x^4 + 1$. For more details and implementation please refer to the application note "CRC Checksum Calculation for SHT2x".

[15] This status bit is updated after each measurement

5.8 Serial Number

SHT21 provides an electronic identification code. For instructions on how to read the identification code please refer to the Application Note "Electronic Identification Code" – to be downloaded from the web page www.sensirion.com/SHT21.

6 Conversion of Signal Output

Default resolution is set to 12 bit relative humidity and 14 bit temperature reading. Measured data are transferred in two byte packages, i.e. in frames of 8 bit length where the most significant bit (MSB) is transferred first (left aligned). Each byte is followed by an acknowledge bit. The two status bits, the last bits of LSB, must be set to '0' before calculating physical values. In the example of Figure 15 and Figure 16, the transferred 16 bit relative humidity data is '0110'0011'0101'0000' = 25424.

6.1 Relative Humidity Conversion

With the relative humidity signal output S_{RH} the relative humidity RH is obtained by the following formula (result in %RH), no matter which resolution is chosen:

$$RH = -6 + 125 \cdot \frac{S_{RH}}{2^{16}}$$

In the example given in Figure 15 and Figure 16 the relative humidity results to be 42.5%RH.

The physical value RH given above corresponds to the relative humidity above liquid water according to World Meteorological Organization (WMO). For relative humidity above ice RH_i the values need to be transformed from relative humidity above water RH_w at temperature t. The equation is given in the following, compare also Application Note "Introduction to Humidity":

$$RH_i = RH_w \cdot \exp\left(\frac{\beta_w \cdot t}{\lambda_w + t}\right) \Big/ \exp\left(\frac{\beta_i \cdot t}{\lambda_i + t}\right)$$

Units are %RH for relative humidity and °C for temperature. The corresponding coefficients are defined as follows: β_w = 17.62, λ_w = 243.12°C, β_i = 22.46, λ_i = 272.62°C.

6.2 Temperature Conversion

The temperature T is calculated by inserting temperature signal output S_T into the following formula (result in °C), no matter which resolution is chosen:

$$T = -46.85 + 175.72 \cdot \frac{S_T}{2^{16}}$$

7 Environmental Stability

The SHT2x sensor series were tested based on AEC-Q100 Rev. G qualification test method where applicable. Sensor specifications are tested to prevail under the AEC-Q100 temperature grade 1 test conditions listed in Table 9[16].

Environment	Standard	Results[17]
HTOL	125°C, 408 hours	Pass
TC	-50°C - 125°C, 1000 cycles	Pass
UHST	130°C / 85%RH / ≈2.3bar, 96h	Pass
THB	85°C / 85%RH, 1000h	Pass
HTSL	150°C, 1000h	Pass
ELFR	125°C, 48h	Pass
ESD immunity	HBM ±4kV, MM ±200V, CDM 750V/500V (corner/other pins)	Pass
Latch-up	force current of ±100mA with T_{amb} = 125°C	Pass

Table 9: Performed qualification test series. HTOL = High Temperature Operating Lifetime, TC = Temperature Cycles, UHST = Unbiased Highly accelerated Stress Test, THB = Temperature Humidity Biased, HTSL = High Temperature Storage Lifetime, ELFR = Early Life Failure Rate. For details on ESD see Sect. 4.1.

Sensor performance under other test conditions cannot be guaranteed and is not part of the sensor specifications. Especially, no guarantee can be given for sensor performance in the field or for customer's specific application.

If sensors are qualified for reliability and behavior in extreme conditions, please make sure that they experience same conditions as the reference sensor. It should be taken into account that response times in assemblies may be longer, hence enough dwell time for the measurement shall be granted. For detailed information please consult Application Note "Testing Guide".

8 Packaging

8.1 Packaging Type

SHT2x sensors are provided in DFN packaging (in analogy with QFN packaging). DFN stands for Dual Flat No leads.

The sensor chip is mounted to a lead frame made of Cu and plated with Ni/Pd/Au. Chip and lead frame are over molded by green epoxy-based mold compound. Please note that side walls of sensors are diced and hence lead

[16] Temperature range is -40 to 125°C (AEC-Q100 temperature grade 1).
[17] According to accuracy and long term drift specification given on Page 2.

frame at diced edge is not covered with respective protective coating. The total weight of the sensor is 25mg.

8.2 Filter Cap and Sockets

For SHT2x a filter cap SF2 will is available. It is designed for fast response times and compact size. Please find the datasheet on Sensirion's web page.

For testing of SHT2x sensors sockets, such as from Plastronics, part number 10LQ50S13030 are recommended (see e.g. www.locknest.com).

8.3 Traceability Information

All SHT2x are laser marked with an alphanumeric, five-digit code on the sensor – see Figure 19.

The marking on the sensor consists of two lines with five digits each. The first line denotes the sensor type (SHT21). The first digit of the second line defines the output mode (D = digital, Sensibus and I²C, P = PWM, S = SDM). The second digit defines the manufacturing year (0 = 2010, 1 = 2011, etc.). The last three digits represent an alphanumeric tracking code. That code can be decoded by Sensirion only and allows for tracking on batch level through production, calibration and testing – and will be provided upon justified request.

Figure 19 Laser marking on SHT21. For details see text.

Reels are also labeled, as displayed in Figure 20 and Figure 21, and give additional traceability information.

Lot No.: XXO-NN-YRRRTTTTT
Quantity: RRRR
RoHS: Compliant

Lot No.

Figure 20: First label on reel: XX = Sensor Type (21 for SHT21), O = Output mode (D = Digital, P = PWM, S = SDM), NN = product revision no., Y = last digit of year, RRR = number of sensors on reel divided by 10 (200 for 2000 units), TTTTT = Traceability Code.

SENSIRION
THE SENSOR COMPANY

Device Type:	1-100PPP-NN
Description:	Humidity & Temperature Sensor SHTxx
Part Order No.	1-100PPP-NN or Customer Number
Date of Delivery:	DD.MM.YYYY
Order Code:	46CCCC / 0

Figure 21: Second label on reel: For Device Type and Part Order Number (See Packaging Information on page 2), Delivery Date (also Date Code) is date of packaging of sensors (DD = day, MM = month, YYYY = year), CCCC = Sensirion order number.

8.4 Shipping Package

SHT2x are provided in tape & reel shipment packaging, sealed into antistatic ESD bags. Standard packaging sizes are 400, 1500 and 5000 units per reel. For SHT21, each reel contains 440mm (55 pockets) header tape and 200mm (25 pockets) trailer tape.

The drawing of the packaging tapes with sensor orientation is shown in Figure 22. The reels are provided in sealed antistatic bags.

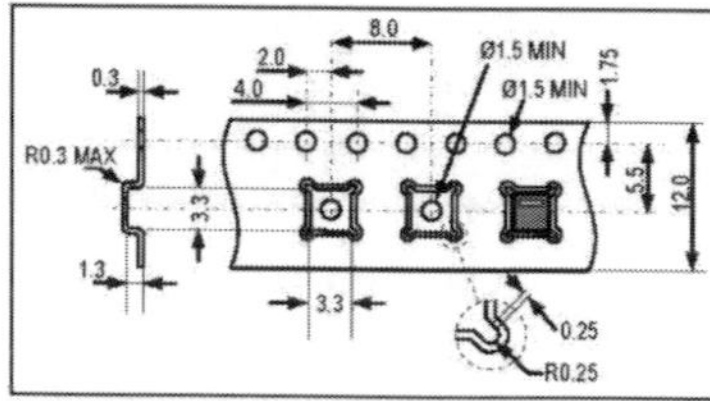

Figure 22 Sketch of packaging tape and sensor orientation. Header tape is to the right and trailer tape to the left on this sketch.

9 Compatibility to SHT1x / 7x protocol

SHT2x sensors may be run by communicating with the Sensirion specific communication protocol used for SHT1x and SHT7x. In case such protocol is applied please refer to the communication chapter of datasheet SHT1x or SHT7x. Please note that reserved status bits of user register must not be changed.

Please understand that with the SHT1x/7x communication protocol only functions described in respective datasheets can be used with the exception of the OTP Reload function that is not set to default on SHT2x. As an alternative to OTP Reload the soft reset may be used.

Please note that even if SHT1x/7x protocol is applied the timing values of Table 5 and Table 7 in this SHT2x datasheet apply.

For the calculation of physical values the following equation must be applied:

For relative humidity *RH*

$$RH = -6 + 125 \cdot \frac{S_{RH}}{2^{RES}}$$

and for temperature *T*

$$T = -46.85 + 175.72 \cdot \frac{S_T}{2^{RES}}$$

RES is the chosen respective resolution, e.g. 12 (12bit) for relative humidity and 14 (14bit) for temperature.

Revision History

Date	Version	Page(s)	Changes
6 May 2009	0.3	1 – 9	Initial preliminary release
21 January 2010	1.0	1 – 4, 7 – 10	Complete revision. For complete revision list please require respective document.
5 May 2010	1.1	1 – 12	Typical specification for temperature sensor. Elimination of errors. For detailed information, please require complete change list at info@sensirion.com.
9 May 2011	2	1 – 7, 10 – 13	Updated temperature accuracy specifications, MSL and standards. Elimination of errors. For detailed information, please require complete change list at info@sensirion.com.
December 2011	3	1, 7-10	Tolerance of threshold value for low battery signal, minor text adaptations and corrections.
May 2014	4	1-4, 7-8, 9-10	Sensor window dimension updated, several minor adjustments

Important Notices

Do not use this product as safety or emergency stop devices or in any other application where failure of the product could result in personal injury. Do not use this product for applications other than its intended and authorized use. Before installing, handling, using or servicing this product, please consult the data sheet and application notes. Failure to comply with these instructions could result in death or serious injury.

If the Buyer shall purchase or use SENSIRION products for any unintended or unauthorized application, Buyer shall defend, indemnify and hold harmless SENSIRION and its officers, employees, subsidiaries, affiliates and distributors against all claims, costs, damages and expenses, and reasonable attorney fees arising out of, directly or indirectly, any claim of personal injury or death associated with such unintended or unauthorized use, even if SENSIRION shall be allegedly negligent with respect to the design or the manufacture of the product.

ESD Precautions

The inherent design of this component causes it to be sensitive to electrostatic discharge (ESD). To prevent ESD-induced damage and/or degradation, take customary and statutory ESD precautions when handling this product.
See application note "ESD, Latchup and EMC" for more information.

Warranty

SENSIRION warrants solely to the original purchaser of this product for a period of 12 months (one year) from the date of delivery that this product shall be of the quality, material and workmanship defined in SENSIRION's published specifications of the product. Within such period, if proven to be defective, SENSIRION shall repair and/or replace this product, in SENSIRION's discretion, free of charge to the Buyer, provided that:

- notice in writing describing the defects shall be given to SENSIRION within fourteen (14) days after their appearance;
- such defects shall be found, to SENSIRION's reasonable satisfaction, to have arisen from SENSIRION's faulty design, material, or workmanship;
- the defective product shall be returned to SENSIRION's factory at the Buyer's expense; and
- the warranty period for any repaired or replaced product shall be limited to the unexpired portion of the original period.

This warranty does not apply to any equipment which has not been installed and used within the specifications recommended by SENSIRION for the intended and proper use of the equipment. EXCEPT FOR THE WARRANTIES EXPRESSLY SET FORTH HEREIN, SENSIRION MAKES NO WARRANTIES, EITHER EXPRESS OR IMPLIED, WITH RESPECT TO THE PRODUCT. ANY AND ALL WARRANTIES, INCLUDING WITHOUT LIMITATION, WARRANTIES OF MERCHANTABILITY OR FITNESS FOR A PARTICULAR PURPOSE, ARE EXPRESSLY EXCLUDED AND DECLINED.

SENSIRION is only liable for defects of this product arising under the conditions of operation provided for in the data sheet and proper use of the goods. SENSIRION explicitly disclaims all warranties, express or implied, for any period during which the goods are operated or stored not in accordance with the technical specifications.

SENSIRION does not assume any liability arising out of any application or use of any product or circuit and specifically disclaims any and all liability, including without limitation consequential or incidental damages. All operating parameters, including without limitation recommended parameters, must be validated for each customer's applications by customer's technical experts. Recommended parameters can and do vary in different applications.

SENSIRION reserves the right, without further notice, (i) to change the product specifications and/or the information in this document and (ii) to improve reliability, functions and design of this product.

Headquarters and Subsidiaries

SENSIRION AG
Laubisruetistr. 50
CH-8712 Staefa ZH
Switzerland

phone: +41 44 306 40 00
fax: +41 44 306 40 30
info@sensirion.com
www.sensirion.com

Sensirion AG (Germany)
phone: +41 44 927 11 66
info@sensirion.com
www.sensirion.com

Sensirion Inc., USA
phone: +1 805 409 4900
info_us@sensirion.com
www.sensirion.com

Sensirion Japan Co. Ltd.
phone: +81 3 3444 4940
info@sensirion.co.jp
www.sensirion.co.jp

Sensirion Korea Co. Ltd.
phone: +82 31 337 7700~3
info@sensirion.co.kr
www.sensirion.co.kr

Sensirion China Co. Ltd.
phone: +86 755 8252 1501
info@sensirion.com.cn
www.sensirion.com.cn

To find your local representative, please visit www.sensirion.com/contact

資料來源：Sensirion 官網：https://www.sensirion.com/fileadmin/user_upload/customers/sensirion/Dokumente/Humidity_Sensors/Sensirion_Humidity_Sensors_SHT21_Datasheet_V4.pdf

SHT31 Datasheet

Datasheet SHT3x-DIS

Humidity and Temperature Sensor

- Fully calibrated, linearized, and temperature compensated digital output
- Wide supply voltage range, from 2.4 V to 5.5 V
- I2C Interface with communication speeds up to 1 MHz and two user selectable addresses
- Typical accuracy of ± 1.5 %RH and ± 0.2 °C for SHT35
- Very fast start-up and measurement time
- Tiny 8-Pin DFN package

Product Summary

SHT3x-DIS is the next generation of Sensirion's temperature and humidity sensors. It builds on a new CMOSens® sensor chip that is at the heart of Sensirion's new humidity and temperature platform. The SHT3x-DIS has increased intelligence, reliability and improved accuracy specifications compared to its predecessor. Its functionality includes enhanced signal processing, two distinctive and user selectable I2C addresses and communication speeds of up to 1 MHz. The DFN package has a footprint of 2.5 x 2.5 mm² while keeping a height of 0.9 mm. This allows for integration of the SHT3x-DIS into a great variety of applications. Additionally, the wide supply voltage range of 2.4 V to 5.5 V guarantees compatibility with diverse assembly situations. All in all, the SHT3x-DIS incorporates 15 years of knowledge of Sensirion, the leader in the humidity sensor industry.

Benefits of Sensirion's CMOSens® Technology

- High reliability and long-term stability
- Industry-proven technology with a track record of more than 15 years
- Designed for mass production
- High process capability
- High signal-to-noise ratio

Content

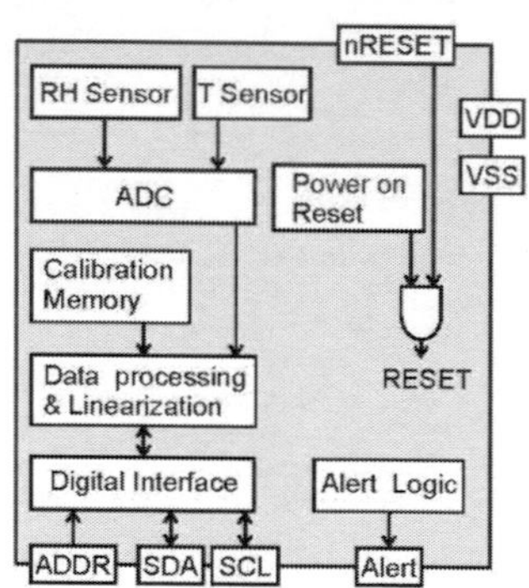

Figure 1 Functional block diagram of the SHT3x-DIS. The sensor signals for humidity and temperature are factory calibrated, linearized and compensated for temperature and supply voltage dependencies.

1 Sensor Performance

Humidity Sensor Specification

Parameter	Condition	Value	Units
SHT30 Accuracy tolerance[1]	Typ.	±3	%RH
	Max.	**Figure 2**	-
SHT31 Accuracy tolerance[1]	Typ.	±2	%RH
	Max.	**Figure 3**	-
SHT35 Accuracy tolerance[1]	Typ.	±1.5	%RH
	Max.	**Figure 4**	-
Repeatability[2]	Low	0.25	%RH
	Medium	0.15	%RH
	High	0.10	%RH
Resolution	Typ.	0.01	%RH
Hysteresis	at 25°C	±0.8	%RH
Specified range[3]	extended[4]	0 to 100	%RH
Response time[5]	$\tau_{63\%}$	8[6]	s
Long-term drift	Typ.[7]	<0.25	%RH/yr

Table 1 Humidity sensor specification.

Temperature Sensor Specification

Parameter	Condition	Value	Units
SHT30 Accuracy tolerance[1]	typ., 0°C to 65°C	±0.3	°C
SHT31 Accuracy tolerance[1]	typ., -40°C to 90°C	±0.3	°C
SHT35 Accuracy tolerance[1]	typ., -40°C to 90°C	±0.2	°C
Repeatability[2]	Low	0.24	°C
	Medium	0.12	°C
	High	0.08	°C
Resolution	Typ.	0.015	°C
Specified Range	-	-40 to 125	°C
Response time [8]	$\tau_{63\%}$	>2	s
Long Term Drift	max	<0.03	°C/yr

Table 2 Temperature sensor specification.

[1] For definition of typical and maximum accuracy tolerance, please refer to the document "Sensirion Humidity Sensor Specification Statement".

[2] The stated repeatability is 3 times the standard deviation (3σ) of multiple consecutive measurements at the stated repeatability and at constant ambient conditions. It is a measure for the noise on the physical sensor output. Different measurement modes allow for high/medium/low repeatability.

[3] Specified range refers to the range for which the humidity or temperature sensor specification is guaranteed.

[4] For details about recommended humidity and temperature operating range, please refer to section 1.1.

[5] Time for achieving 63% of a humidity step function, valid at 25°C and 1m/s airflow. Humidity response time in the application depends on the design-in of the sensor.

[6] With activated ART function (see section 4.7) the response time can be improved by a factor of 2.

[7] Typical value for operation in normal RH/T operating range, see section 1.1. Maximum value is < 0.5 %RH/yr. Higher drift values might occur due to contaminant environments with vaporized solvents, out-gassing tapes, adhesives, packaging materials, etc. For more details please refer to Handling Instructions.

[8] Temperature response times strongly depend on the type of heat exchange, the available sensor surface and the design environment of the sensor in the final application.

Humidity Sensor Performance Graphs

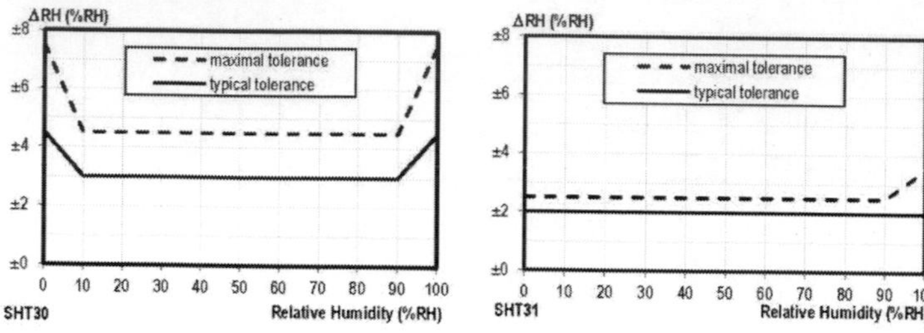

Figure 2 Tolerance of RH at 25°C for SHT30.

Figure 3 Tolerance of RH at 25°C for SHT31.

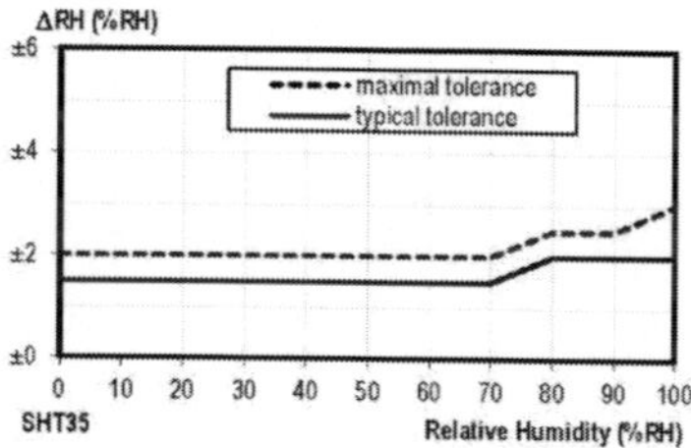

Figure 4 Tolerance of RH at 25°C for SHT35.

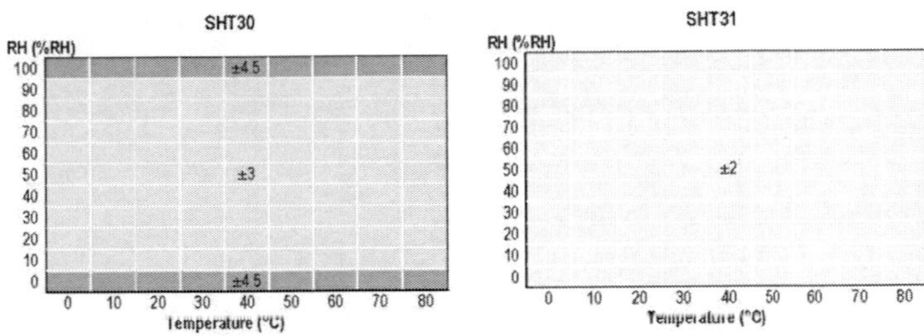

Figure 5 Typical tolerance of RH over T for SHT30.

Figure 6 Typical tolerance of RH over T for SHT31.

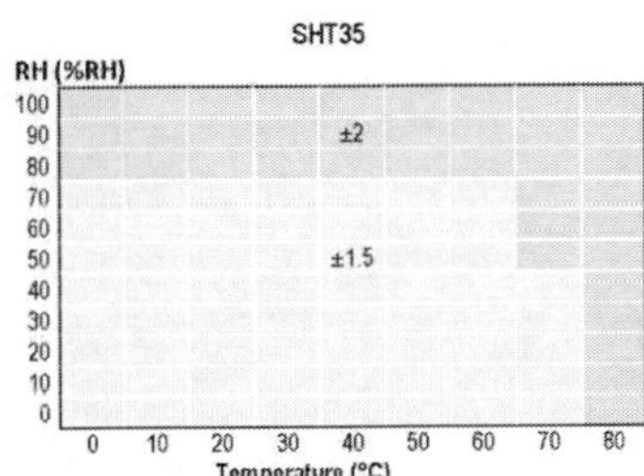

Figure 7 Typical tolerance of RH over T for SHT35.

Temperature Sensor Performance Graphs

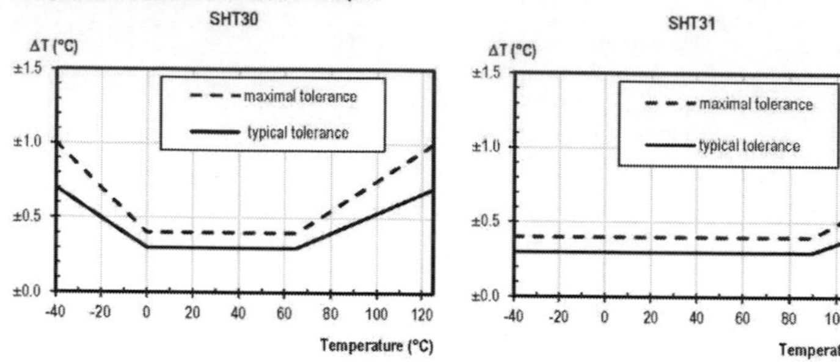

Figure 8 Temperature accuracy of the SHT30 sensor.

Figure 9 Temperature accuracy of the SHT31 sensor.

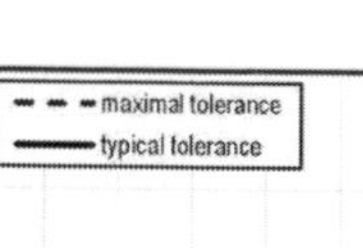

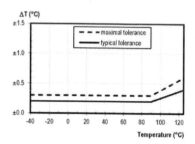

Figure 10 Temperature accuracy of the SHT35 sensor.

1.1 Recommended Operating Condition

The sensor shows best performance when operated within recommended normal temperature and humidity range of 5 °C – 60 °C and 20 %RH – 80 %RH, respectively. Long-term exposure to conditions outside normal range, especially at high humidity, may temporarily offset the RH signal (e.g. +3%RH after 60h kept at >80%RH). After returning into the normal temperature and humidity range the sensor will slowly come back to calibration state by itself. Prolonged exposure to extreme conditions may accelerate ageing. To ensure stable operation of the humidity sensor, the conditions described in the document "SHTxx Assembly of SMD Packages", section "Storage and Handling Instructions" regarding exposure to volatile organic compounds have to be met. Please note as well that this does apply not only to transportation and manufacturing, but also to operation of the SHT3x-DIS.

2 Specifications

2.1 Electrical Specifications

Parameter	Symbol	Condition	Min.	Typ.	Max.	Units	Comments
Supply voltage	V_{DD}		2.4	3.3	5.5	V	
Power-up/down level	V_{POR}		2.1	2.3	2.4	V	
Slew rate change of the supply voltage	$V_{DD,slew}$		-	-	20	V/ms	Voltage changes on the VDD line between $V_{DD,min}$ and $V_{DD,max}$ should be slower than the maximum slew rate; faster slew rates may lead to reset;
Supply current	I_{DD}	idle state (single shot mode)	-	0.2	2.0	µA	Current when sensor is not performing a measurement during single shot mode
		idle state (periodic data acquisition mode)	-	45	70	µA	Current when sensor is not performing a measurement during periodic data acquisition mode
		Measuring	-	800	1500	µA	Current consumption while sensor is measuring
		Average	-	2	5	µA	Current consumption (operation with one measurement per second at lowest repeatability, single shot mode)
Alert Output driving strength	IOH		0.8x V_{DD}	1.5x V_{DD}	2.1x V_{DD}	mA	See also section 3.5
Heater power	P_{Heater}	Heater running	5	-	25	mW	Depending on the supply voltage

Table 3 Electrical specifications, valid at 25°C.

2.2 Timing Specification for the Sensor System

Parameter	Symbol	Conditions	Min.	Typ.	Max.	Units	Comments
Power-up time	t_{PU}	After hard reset, $V_{DD} \geq V_{POR}$	-	0.5	1	ms	Time between V_{DD} reaching V_{POR} and sensor entering idle state
Soft reset time	t_{SR}	After soft reset.	-	0.5	1	ms	Time between ACK of soft reset command and sensor entering idle state
Duration of reset pulse	t_{RESETN}		350	-	-	ns	See section 3.6
Measurement duration	$t_{MEAS,l}$	Low repeatability	-	2.5	4	ms	The three repeatability modes differ with respect to measurement duration, noise level and energy consumption.
	$t_{MEAS,m}$	Medium repeatability	-	4.5	6	ms	
	$t_{MEAS,h}$	High repeatability	-	12.5	15	ms	

Table 4 System timing specification, valid from -40 °C to 125 °C and 2.4 V to 5.5 V.

2.3 Absolute Minimum and Maximum Ratings

Stress levels beyond those listed in Table 5 may cause permanent damage to the device or affect the reliability of the sensor. These are stress ratings only and functional operation of the device at these conditions is not guaranteed.

Parameter	Rating	Units
Supply voltage V_{DD}	-0.3 to 6	V
Max Voltage on pins (pin 1 (SDA); pin 2 (ADDR); pin 3 (ALERT); pin 4 (SCL); pin 6 (nRESET))	-0.3 to VDD+0.3	V
Input current on any pin	±100	mA
Operating temperature range	-40 to 125	°C
Storage temperature range	-40 to 150	°C
ESD HBM (human body model)[9]	4	kV
ESD CDM (charge device model)[10]	750	V

Table 5 Minimum and maximum ratings; voltage values may only be applied for short time periods.

[9] According to JEDEC JS-001-2014, classification level 3A.

[10] According to JEDEC JS-002-2014, classification level C2b.

3 Pin Assignment

The SHT3x-DIS comes in a tiny 8-pin DFN package – see Table 6.

Pin	Name	Comments
1	SDA	Serial data; input / output
2	ADDR	Address pin; input; connect to either VDD or VSS using a series resistor of ~1 kΩ, do not leave floating
3	ALERT	Indicates alarm condition; output; must be left floating if unused
4	SCL	Serial clock; input / output
5	VDD	Supply voltage; input
6	nRESET	Reset pin active low; input; if not used must be left floating
7	R	No electrical function; to be connected to VSS
8	VSS	Ground

Table 6 SHT3x-DIS pin assignment (transparent top view). Dashed lines are only visible if viewed from below. The die pad is internally connected to VSS.

3.1 Power Pins (VDD, VSS)

The electrical specifications of the SHT3x-DIS are shown in Table 3. The power supply pins must be decoupled with a 100 nF capacitor that shall be placed as close to the sensor as possible – see Figure 11 for a typical application circuit.

3.2 Serial Clock and Serial Data (SCL, SDA)

SCL is used to synchronize the communication between microcontroller and the sensor. The clock frequency can be freely chosen between 0 to 1000 kHz. Commands with clock stretching according to I2C Standard[11] are supported.

The SDA pin is used to transfer data to and from the sensor. Communication with frequencies up to 400 kHz must meet the I2C *Fast Mode*[11] standard. Communication frequencies up to 1 Mhz are supported following the specifications given in **Table 20**.

Both SCL and SDA lines are open-drain I/Os with diodes to VDD and VSS. They should be connected to external pull-up resistors (please refer to Figure 11). A device on the I2C bus must only drive a line to ground. The external pull-up resistors (e.g. R_p=10 kΩ) are required to pull the signal high. For dimensioning resistor sizes please take bus capacity and communication frequency into account (see for example Section 7.1 of NXPs I2C Manual for more details[11]). It should be noted that pull-up resistors may be included in I/O circuits of microcontrollers. It is recommended to wire the sensor according to the application circuit as shown in Figure 11.

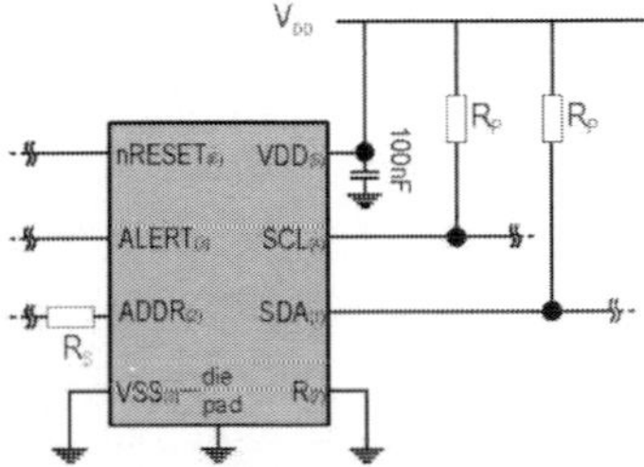

Figure 11 Typical application circuit. Please note that the positioning of the pins does not reflect the position on the real sensor. This is shown in Table 6.

3.3 Die Pad (center pad)

The die pad or center pad is visible from below and located in the center of the package. It is electrically connected to VSS. Hence electrical considerations do not impose constraints on the wiring of the die pad. However, due to mechanical reasons it is recommended to solder the center pad to the PCB. For more information on design-in, please refer to the document "SHTxx Design Guide".

3.4 ADDR Pin

Through the appropriate wiring of the ADDR pin the I2C address can be selected (see Table 7 for the respective addresses). Aside from hard wiring the ADDR pin to VDD or VSS, it is also possible to use it as a selector pin. This means that the address of the sensor can be changed dynamically during operation by switching the level on the ADDR pin. The only constraint is that the level has to stay constant starting from the I2C start condition until the communication is finished. This allows to connect more than two SHT3x-DIS onto the same bus. The dynamical switching requires individual ADDR lines to the sensors.

[11] http://www.nxp.com/documents/user_manual/UM10204.pdf

Please note that the I2C address is represented through the 7 MSBs of the I2C read or write header. The LSB switches between read or write header. The wiring for the default address is shown in Table 7 and Figure 11. The ADDR pin must not be left floating. Please note that only the 7 MSBs of the I2C Read/Write header constitute the I2C Address.

SHT3x-DIS	I2C Address in Hex. representation	Condition
I2C address A	0x44 (default)	ADDR (pin 2) connected to VSS
I2C address B	0x45	ADDR (pin 2) connected to VDD

Table 7 I2C device addresses.

3.5 ALERT Pin

The alert pin may be used to connect to the interrupt pin of a microcontroller. The output of the pin depends on the value of the RH/T reading relative to programmable limits. Its function is explained in a separate application note. If not used, this pin must be left floating. The pin switches high, when alert conditions are met. The maximum driving loads are listed in Table 3. Be aware that self-heating might occur, depending on the amount of current that flows. Self-heating can be prevented if the Alert Pin is only used to switch a transistor.

3.6 nRESET Pin

The nReset pin may be used to generate a reset of the sensor. A minimum pulse duration of 350 ns is required to reliably trigger a reset of the sensor. Its function is explained in more detail in section 4. If not used it must be left floating.

4 Operation and Communication

The SHT3x-DIS supports I2C fast mode (and frequencies up to 1000 kHz). Clock stretching can be enabled and disabled through the appropriate user command. For detailed information on the I2C protocol, refer to NXP I2C-bus specification[12].

All SHT3x-DIS commands and data are mapped to a 16-bit address space. Additionally, data and commands are protected with a CRC checksum. This increases communication reliability. The 16 bits commands to the sensor already include a 3 bit CRC checksum. Data sent from and received by the sensor is always succeeded by an 8 bit CRC.

In write direction it is mandatory to transmit the checksum, since the SHT3x-DIS only accepts data if it is followed by the correct checksum. In read direction it is left to the master to read and process the checksum.

4.1 Power-Up and Communication Start

The sensor starts powering-up after reaching the power-up threshold voltage V_{POR} specified in Table 3. After reaching this threshold voltage the sensor needs the time t_{PU} to enter idle state. Once the idle state is entered it is ready to receive commands from the master (microcontroller).

Each transmission sequence begins with a START condition (S) and ends with a STOP condition (P) as described in the I2C-bus specification. The stop condition is optional. Whenever the sensor is powered up, but not performing a measurement or communicating, it automatically enters idle state for energy saving. This idle state cannot be controlled by the user.

4.2 Starting a Measurement

A measurement communication sequence consists of a START condition, the I2C write header (7-bit I2C device address plus 0 as the write bit) and a 16-bit measurement command. The proper reception of each byte is indicated by the sensor. It pulls the SDA pin low (ACK bit) after the falling edge of the 8th SCL clock to indicate the reception. A complete measurement cycle is depicted in Table 8.

With the acknowledgement of the measurement command, the SHT3x-DIS starts measuring humidity and temperature.

4.3 Measurement Commands for Single Shot Data Acquisition Mode

In this mode one issued measurement command triggers the acquisition of *one data pair*. Each data pair consists of one 16 bit temperature and one 16 bit humidity value (in this order). During transmission each data value is always followed by a CRC checksum, see section 4.4.

In single shot mode different measurement commands can be selected. The 16 bit commands are shown in Table 8. They differ with respect to repeatability (low, medium and high) and clock stretching (enabled or disabled).

The repeatability setting influences the measurement duration and thus the overall energy consumption of the sensor. This is explained in section 0.

[12] http://www.nxp.com/documents/user_manual/UM10204.pdf

Condition		Hex. code	
Repeatability	Clock stretching	MSB	LSB
High	enabled	0x2C	06
Medium			0D
Low			10
High	disabled	0x24	00
Medium			0B
Low			16

e.g. 0x2C06: high repeatability measurement with clock stretching enabled

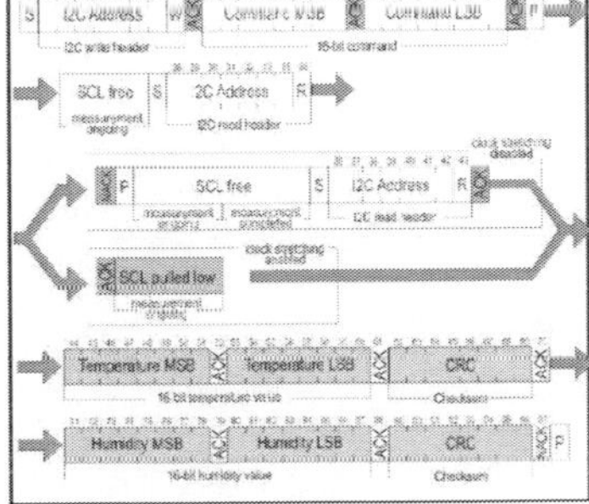

Table 8 Measurement commands in single shot mode (Clear blocks are controlled by the microcontroller, grey blocks by the sensor).

4.4 Readout of Measurement Results for Single Shot Mode

After the sensor has completed the measurement, the master can read the measurement results (pair of RH& T) by sending a START condition followed by an I2C read header. The sensor will acknowledge the reception of the read header and send two bytes of data (temperature) followed by one byte CRC checksum and another two bytes of data (relative humidity) followed by one byte CRC checksum. Each byte must be acknowledged by the microcontroller with an ACK condition for the sensor to continue sending data. If the sensor does not receive an ACK from the master after any byte of data, it will not continue sending data.

The sensor will send the temperature value first and then the relative humidity value. After having received the checksum for the humidity value a NACK and stop condition should be sent (see Table 8).

The I2C master can abort the read transfer with a NACK condition after any data byte if it is not interested in subsequent data, e.g. the CRC byte or the second measurement result, in order to save time.

In case the user needs humidity and temperature data but does not want to process CRC data, it is recommended to read the two temperature bytes of data with the CRC byte (without processing the CRC data); after having read the two humidity bytes, the read transfer can be aborted with a with a NACK.

No Clock Stretching

When a command without clock stretching has been issued, the sensor responds to a read header with a not acknowledge (NACK), if no data is present.

Clock Stretching

When a command with clock stretching has been issued, the sensor responds to a read header with an ACK and subsequently pulls down the SCL line. The SCL line is pulled down until the measurement is complete. As soon as the measurement is complete, the sensor releases the SCL line and sends the measurement results.

4.5 Measurement Commands for Periodic Data Acquisition Mode

In this mode one issued measurement command yields *a stream of data pairs*. Each data pair consists of one 16 bit temperature and one 16 bit humidity value (in this order).

In periodic mode different measurement commands can be selected. The corresponding 16 bit commands are shown in Table 9. They differ with respect to repeatability (low, medium and high) and data acquisition frequency (0.5, 1, 2, 4 & 10 measurements per second, mps). Clock stretching cannot be selected in this mode.

The data acquisition frequency and the repeatability setting influences the measurement duration and the current consumption of the sensor. This is explained in section 2 of this datasheet.

If a measurement command is issued, while the sensor is busy with a measurement (measurement durations see Table 4), it is recommended to issue a break command first (see section 4.8). Upon reception of the break command the sensor will finish the ongoing measurement and enter the single shot mode.

Condition		Hex. code	
Repeatability	mps	MSB	LSB
High	0.5	0x20	32
Medium			24
Low			2F
High	1	0x21	30
Medium			26
Low			2D
High	2	0x22	36
Medium			20
Low			2B
High	4	0x23	34
Medium			22
Low			29
High	10	0x27	37
Medium			21
Low			2A
e.g. 0x2130: 1 high repeatability mps - measurement per second			

S | I2C Address | W | ACK | Command MSB | ACK | Command LSB | ACK
I2C write header — 16-bit command

Table 9 Measurement commands for periodic data acquisition mode (Clear blocks are controlled by the microcontroller, grey blocks by the sensor). N.B.: At the highest mps setting self-heating of the sensor might occur.

4.6 Readout of Measurement Results for Periodic Mode

Transmission of the measurement data can be initiated through the fetch data command shown in Table 10. If no measurement data is present the I2C read header is responded with a NACK (Bit 9 in Table 10) and the communication stops. After the read out command fetch data has been issued, the data memory is cleared, i.e. no measurement data is present.

Command	Hex code
Fetch Data	0x E0 00

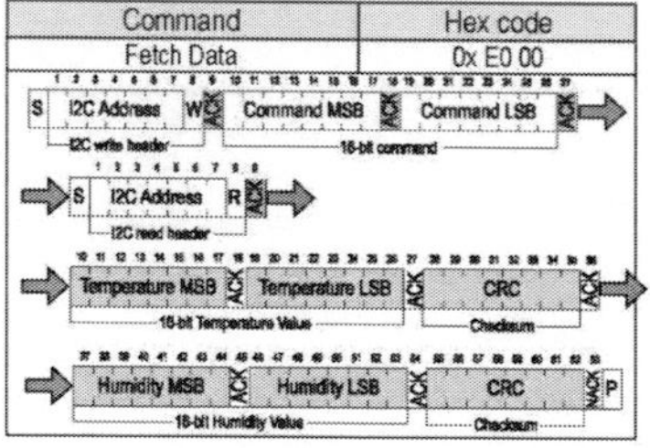

Table 10 Fetch Data command (Clear blocks are controlled by the microcontroller, grey blocks by the sensor).

4.7 ART Command

The ART (accelerated response time) feature can be activated by issuing the command in Table 11. After issuing the ART command the sensor will start acquiring data with a frequency of 4Hz.

The ART command is structurally similar to any other command in Table 9. Hence section 4.5 applies for starting a measurement, section 4.6 for reading out data and section 4.8 for stopping the periodic data acquisition.

The ART feature can also be evaluated using the Evaluation Kit EK-H5 from Sensirion.

Command	Hex Code
Periodic Measurement with ART	0x2B32

S | I2C Address | W | ACK | Command MSB | ACK | Command LSB | ACK
I2C write header — 16-bit command

Table 11 Command for a periodic data acquisition with the ART feature (Clear blocks are controlled by the microcontroller, grey blocks by the sensor).

4.8 Break command / Stop Periodic Data Acquisition Mode

The periodic data acquisition mode can be stopped using the break command shown in Table 12. It is recommended to stop the periodic data acquisition prior to sending another command (except Fetch Data command) using the break command. Upon reception of the break command the sensor enters the single shot mode, after finishing the ongoing measurement. This can take up to 15 ms, depending on the selected repeatability.

Command	Hex Code
Break	0x3093

S | I2C Address | W | ACK | Command MSB | ACK | Command LSB | ACK | P
I2C write header — 16-bit command

Table 12 Break command (Clear blocks are controlled by the microcontroller, grey blocks by the sensor).

4.9 Reset

A system reset of the SHT3x-DIS can be generated externally by issuing a command (soft reset) or by sending a pulse to the dedicated reset pin (nReset pin). Additionally, a system reset is generated internally during power-up. During the reset procedure the sensor will not process commands.

In order to achieve a full reset of the sensor without removing the power supply, it is recommended to use the nRESET pin of the SHT3x-DIS.

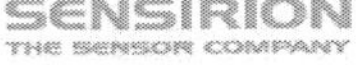

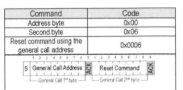

Interface Reset

If communication with the device is lost, the following signal sequence will reset the serial interface: While leaving SDA high, toggle SCL nine or more times. This must be followed by a Transmission Start sequence preceding the next command. This sequence resets the interface only. The status register preserves its content.

Soft Reset / Re-Initialization

The SHT3x-DIS provides a soft reset mechanism that forces the system into a well-defined state without removing the power supply. When the system is in idle state the soft reset command can be sent to the SHT3x-DIS. This triggers the sensor to reset its system controller and reloads calibration data from the memory. In order to start the soft reset procedure the command as shown in Table 13 should be sent.

It is worth noting that the sensor reloads calibration data prior to every measurement by default.

Command	Hex Code
Soft Reset	0x30A2

S | I2C Address | W | ACK | Command MSB | ACK | Command LSB | ACK | P
I2C write header — 16-bit command

Table 13 Soft reset command (Clear blocks are controlled by the microcontroller, grey blocks by the sensor).

Reset through General Call

Additionally, a reset of the sensor can also be generated using the "general call" mode according to I2C-bus specification[12]. This generates a reset which is functionally identical to using the nReset pin. It is important to understand that a reset generated in this way is not device specific. All devices on the same I2C bus that support the general call mode will perform a reset. Additionally, this command only works when the sensor is able to process I2C commands. The appropriate command consists of two bytes and is shown in Table 14.

Reset through the nReset Pin

Pulling the nReset pin low (see Table 6) generates a reset similar to a hard reset. The nReset pin is internally connected to VDD through a pull-up resistor and hence active low. The nReset pin has to be pulled low for a minimum of 350 ns to generate a reset of the sensor.

Command	Code
Address byte	0x00
Second byte	0x06
Reset command using the general call address	0x0006

S | General Call Address | ACK | Reset Command | ACK
General Call 1st byte — General Call 2nd byte

Table 14 Reset through the general call address (Clear blocks are controlled by the microcontroller, grey blocks by the sensor)

Hard Reset

A hard reset is achieved by switching the supply voltage to the VDD Pin off and then on again. In order to prevent powering the sensor over the ESD diodes, the voltage to pins 1 (SDA), 4 (SCL) and 2 (ADDR) also needs to be removed.

4.10 Heater

The heater can be switched on and off by command, see table below. The status is listed in the status register. After a reset the heater is disabled (default condition).

Command	Hex Code	
	MSB	LSB
Heater Enable	0x30	6D
Heater Disabled		66

S | I2C Address | W | ACK | Command MSB | ACK | Command LSB | ACK | P
I2C write header — 16-bit command

Table 15 Heater command (Clear blocks are controlled by the microcontroller, grey blocks by the sensor).

4.11 Status Register

The status register contains information on the operational status of the heater, the alert mode and on the execution status of the last command and the last write sequence. The command to read out the status register is shown in Table 16 whereas a description of the content can be found in Table 17.

Command	Hex code
Read Out of status register	0xF32D

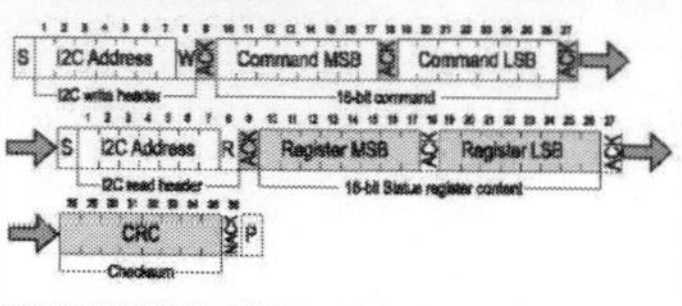

S I2C Address W ACK Command MSB ACK Command LSB ACK
I2C write header — 16-bit command
S I2C Address R ACK Register MSB ACK Register LSB ACK
I2C read header — 16-bit Status register content
CRC ACK P
Checksum

Table 16 Command to read out the status register (Clear blocks are controlled by the microcontroller, grey blocks by the sensor).

Bit	Field description	Default value
15	Alert pending status '0': no pending alerts '1': at least one pending alert	'1'
14	Reserved	'0'
13	Heater status '0' : Heater OFF '1' : Heater ON	'0'
12	Reserved	'0'
11	RH tracking alert '0' : no alert '1' . alert	'0
10	T tracking alert '0' : no alert '1' . alert	'0'
9:5	Reserved	'xxxxx'
4	System reset detected '0': no reset detected since last 'clear status register' command '1': reset detected (hard reset, soft reset command or supply fail)	'1'
3:2	Reserved	'00'
1	Command status '0': last command executed successfully '1': last command not processed. It was either invalid, failed the integrated command checksum	'0'
0	Write data checksum status '0': checksum of last write transfer was correct '1': checksum of last write transfer failed	'0'

Table 17 Description of the status register.

Clear Status Register

All flags (Bit 15, 11, 10, 4) in the status register can be cleared (set to zero) by sending the command shown in Table 18.

Command	Hex Code
Clear status register	0x 30 41

S I2C Address W ACK Command MSB ACK Command LSB ACK P
I2C write header — 16-bit command

Table 18 Command to clear the status register (Clear blocks are controlled by the microcontroller, grey blocks by the sensor).

4.12 Checksum Calculation

The 8-bit CRC checksum transmitted after each data word is generated by a CRC algorithm. Its properties are displayed in Table 19. The CRC covers the contents of the two previously transmitted data bytes. To calculate the checksum only these two previously transmitted data bytes are used.

Property	Value
Name	CRC-8
Width	8 bit
Protected data	read and/or write data
Polynomial	0x31 ($x^8 + x^5 + x^4 + 1$)
Initialization	0xFF
Reflect input	False
Reflect output	False
Final XOR	0x00
Examples	CRC (0xBEEF) = 0x92

Table 19 I2C CRC properties.

4.13 Conversion of Signal Output

Measurement data is always transferred as 16-bit values (unsigned integer). These values are already linearized and compensated for temperature and supply voltage effects. Converting those raw values into a physical scale can be achieved using the following formulas.

Relative humidity conversion formula (result in %RH):

$$RH = 100 \cdot \frac{S_{RH}}{2^{16} - 1}$$

Temperature conversion formula (result in °C & °F):

$$T\,[^{\circ}C] = -45 + 175 \cdot \frac{S_T}{2^{16} - 1}$$

$$T\,[^{\circ}F] = -49 + 315 \cdot \frac{S_T}{2^{16} - 1}$$

S_{RH} and S_T denote the raw sensor output for humidity and temperature, respectively. The formulas work only correctly when S_{RH} and S_T are used in decimal representation.

4.14 Communication Timing

Parameter	Symbol	Conditions	Min.	Typ.	Max.	Units	Comments
SCL clock frequency	f_{SCL}		0	-	1000	kHz	
Hold time (repeated) START condition	$t_{HD;STA}$	After this period, the first clock pulse is generated	0.24	-	-	µs	
LOW period of the SCL clock	t_{LOW}		0.65	-	-	µs	
HIGH period of the SCL clock	t_{HIGH}		0.26	-	-	µs	
SDA hold time	$t_{HD;DAT}$		0	-	250	ns	Transmitting data
			0	-	-	ns	Receiving data
SDA set-up time	$t_{SU;DAT}$		100	-	-	ns	
SCL/SDA rise time	t_R		-	-	300	ns	
SCL/SDA fall time	t_F		-	-	300	ns	
SDA valid time	$t_{VD;DAT}$		-	-	0.9	µs	
Set-up time for a repeated START condition	$t_{SU;STA}$		0.6	-	-	µs	
Set-up time for STOP condition	$t_{SU;STO}$		0.6	-	-	µs	
Capacitive load on bus line	CB		-	-	400	pF	
Low level input voltage	V_{IL}		-0.5	-	$0.3xV_{DD}$	V	
High level input voltage	V_{IH}		$0.7xV_{DD}$	-	$1xV_{DD}$	V	
Low level output voltage	V_{OL}	3 mA sink current	-	-	0.66	V	

Table 20 Communication timing specifications for I2C fm (fast mode), specifications are at 25°C and typical VDD. The numbers above are values according to the I2C Specification (UM10204, Rev. 6, April 4, 2014).

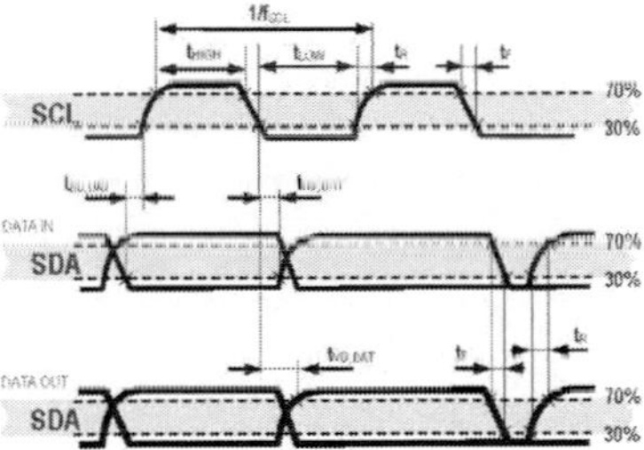

Figure 12 Timing diagram for digital input/output pads. SDA directions are seen from the sensor. Bold SDA lines are controlled by the sensor, plain SDA lines are controlled by the micro-controller. Note that SDA valid read time is triggered by falling edge of preceding toggle.

5 Packaging

SHT3x-DIS sensors are provided in an open-cavity DFN package. DFN stands for dual flat no leads. The humidity sensor opening is centered on the top side of the package.

The sensor chip is made of silicon and is mounted to a lead frame. The latter is made of Cu plated with Ni/Pd/Au. Chip and lead frame are overmolded by an epoxy-based mold compound leaving the central die pad and I/O pins exposed for mechanical and electrical connection. Please note that the side walls of the sensor are diced and therefore these diced lead frame surfaces are not covered with the respective plating.

The package (except for the humidity sensor opening) follows JEDEC publication 95, design registration 4.20, small scale plastic quad and dual inline, square and rectangular, No-LEAD packages (with optional thermal enhancements) small scale (QFN/SON), Issue D.01, September 2009.

SHT3x-DIS has a Moisture Sensitivity Level (MSL) of 1, according to IPC/JEDEC J-STD-020. At the same time, it is recommended to further process the sensors within 1 year after date of delivery.

5.1 Traceability

All SHT3x-DIS sensors are laser marked for easy identification and traceability. The marking on the sensor top side consists of a pin-1 indicator and two lines of text.

The top line consist of the pin-1 indicator which is located in the top left corner and the product name. The small letter x stands for the accuracy class.

The bottom line consists of 6 letters. The first two digits XY (=DI) describe the output mode. The third letter (A) represents the manufacturing year (4 = 2014, 5 = 2015, etc). The last three digits (BCD) represent an alphanumeric tracking code. That code can be decoded by Sensirion only and allows for tracking on batch level through production, calibration and testing – and will be provided upon justified request.

If viewed from below pin 1 is indicated by triangular shaped cut in the otherwise rectangular die pad. The dimensions of the triangular cut are shown in Figure 14 through the labels T1 & T2.

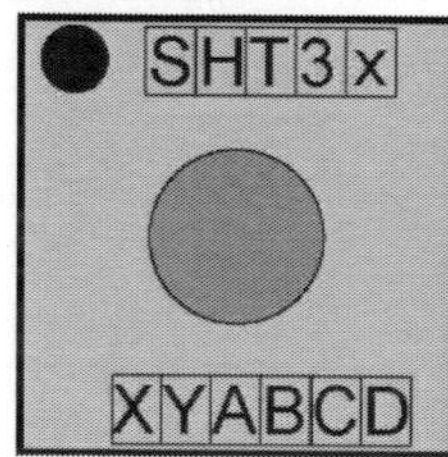

Figure 13 Top view of the SHT3x-DIS illustrating the laser marking.

5.2 Package Outline

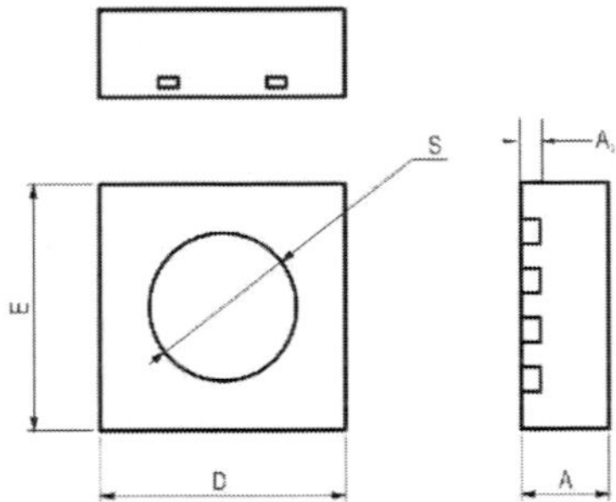

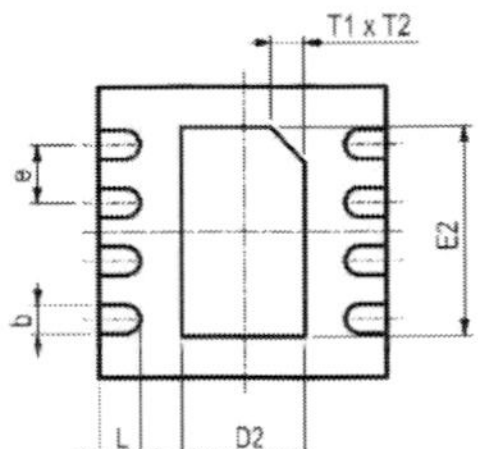

Figure 14 Dimensional drawing of SHT3x-DIS sensor package

Parameter	Symbol	Min	Nom.	Max	Units	Comments
Package height	A	0.8	0.9	1	mm	
Leadframe height	A3	-	0.2	-	mm	
Pad width	b	0.2	0.25	0.3	mm	
Package width	D	2.4	2.5	2.6	mm	
Center pad length	D2	1	1.1	1.2	mm	
Package length	E	2.4	2.5	2.6	mm	
Center pad width	E2	1.7	1.8	1.9	mm	
Pad pitch	e	-	0.5		mm	
Pad length	L	0.3	0.35	0.4	mm	
Max cavity	S	-	-	1.5	mm	Only as guidance. This value includes all tolerances, including displacement tolerances. Typically the opening will be smaller.
Center pad marking	T1xT2	-	0.3x45°	-	mm	indicates the position of pin 1

Table 21 Package outline.

5.3 Land Pattern

Figure 15 shows the land pattern. The land pattern is understood to be the open metal areas on the PCB, onto which the DFN pads are soldered.

The solder mask is understood to be the insulating layer on top of the PCB covering the copper traces. It is recommended to design the solder pads as a Non-Solder Mask Defined (NSMD) type. For NSMD pads, the solder mask opening should provide a 60 µm to 75 µm design clearance between any copper pad and solder mask. As the pad pitch is only 0.5 mm we recommend to have one solder mask opening for all 4 I/O pads on one side.

For solder paste printing it is recommended to use a laser-cut, stainless steel stencil with electro-polished trapezoidal walls and with 0.1 or 0.125 mm stencil thickness. The length of the stencil apertures for the I/O pads should be the same as the PCB pads. However, the position of the stencil apertures should have an offset of 0.1 mm away from the center of the package. The die pad aperture should cover about 70 – 90 % of the die pad area –thus it should have a size of about 0.9 mm x 1.6 mm.

For information on the soldering process and further recommendation on the assembly process please consult the Application Note HT_AN_SHTxx_Assembly_of_SMD_Packages , which can be found on the Sensirion webpage.

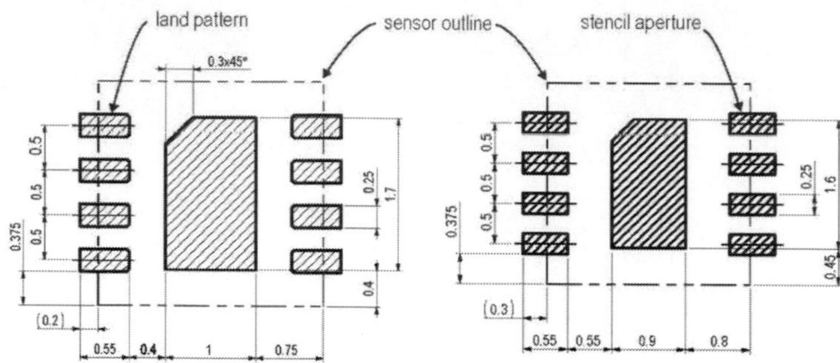

Figure 15 Recommended metal land pattern (left) and stencil apertures (right) for the SHT3x-DIS. The dashed lines represent the outer dimension of the DFN package. The PCB pads (left) and stencil apertures (right) are indicated through the shaded areas.

6 Shipping Package

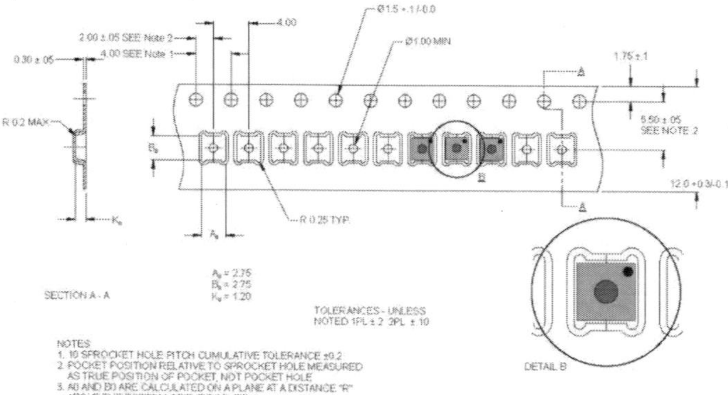

Figure 16 Technical drawing of the packaging tape with sensor orientation in tape. Header tape is to the right and trailer tape to the left on this drawing. Dimensions are given in millimeters.

7 Quality

Qualification of the SHT3x-DIS is performed based on the AEC Q 100 qualification test method.

7.1 Material Contents

The device is fully RoHS and WEEE compliant, e.g. free of Pb, Cd, and Hg.

8 Ordering Information

The SHT3x-DIS can be ordered in tape and reel packaging with different sizes, see Table 22. The reels are sealed into antistatic ESD bags. The document "SHT3x shipping package" that shows the details about the shipping package is available upon request.

Name	Quantity	Order Number
SHT30-DIS-B2.5kS	2500	1-101400-01
SHT30-DIS-B10kS	10000	1-101173-01
SHT31-DIS-B2.5kS	2500	1-101386-01
SHT31-DIS-B10kS	10000	1-101147-01
SHT35-DIS-B2.5kS	2500	1-101388-01
SHT35-DIS-B10kS	10000	1-101479-01

Table 22 SHT3x-DIS ordering options.

9 Further Information

For more in-depth information on the SHT3x-DIS and its application please consult the following documents:

Document Name	Description	Source
SHT3x Shipping Package	Information on Tape, Reel and shipping bags (technical drawing and dimensions)	Available upon request
SHTxx Assembly of SMD Packages	Assembly Guide (Soldering Instructions)	Available for download at the Sensirion humidity sensors download center: www.sensirion.com/humidity-download
SHTxx Design Guide	Design guidelines for designing SHTxx humidity sensors into applications	Available for download at the Sensirion humidity sensors download center: www.sensirion.com/humidity-download
SHTxx Handling Instructions	Guidelines for proper handling of SHTxx humidity sensors	Available for download at the Sensirion humidity sensors download center: www.sensirion.com/humidity-download
Sensirion Humidity Sensor Specification Statement	Definition of sensor specifications.	Available for download at the Sensirion humidity sensors download center: www.sensirion.com/humidity-download

Table 23 Documents containing further information relevant for the SHT3x-DIS.

Revision History

Date	Version	Page(s)	Changes
October 2015	1		-
June 2016	2	2-4	Specifications for SHT35 added
		6	ESD specifications updated
		7	**Table 6** "Comments" section updated
		7	**Figure 11** updated according to **Table 6**
		11	Updated information about data memory to: "After the read out command "fetch data" has been issued, the data memory is reset, i.e. no measurement data is present.
		17	Ordering information in **Table 22** updated

Important Notices

Warning, Personal Injury

Do not use this product as safety or emergency stop devices or in any other application where failure of the product could result in personal injury. Do not use this product for applications other than its intended and authorized use. Before installing, handling, using or servicing this product, please consult the data sheet and application notes. Failure to comply with these instructions could result in death or serious injury.

If the Buyer shall purchase or use SENSIRION products for any unintended or unauthorized application, Buyer shall defend, indemnify and hold harmless SENSIRION and its officers, employees, subsidiaries, affiliates and distributors against all claims, costs, damages and expenses, and reasonable attorney fees arising out of, directly or indirectly, any claim of personal injury or death associated with such unintended or unauthorized use, even if SENSIRION shall be allegedly negligent with respect to the design or the manufacture of the product.

ESD Precautions

The inherent design of this component causes it to be sensitive to electrostatic discharge (ESD). To prevent ESD-induced damage and/or degradation, take customary and statutory ESD precautions when handling this product.
See application note "ESD, Latchup and EMC" for more information.

Warranty

SENSIRION warrants solely to the original purchaser of this product for a period of 12 months (one year) from the date of delivery that this product shall be of the quality, material and workmanship defined in SENSIRION's published specifications of the product. Within such period, if proven to be defective, SENSIRION shall repair and/or replace this product, in SENSIRION's discretion, free of charge to the Buyer, provided that:

- notice in writing describing the defects shall be given to SENSIRION within fourteen (14) days after their appearance;
- such defects shall be found, to SENSIRION's reasonable satisfaction, to have arisen from SENSIRION's faulty design, material, or workmanship;
- the defective product shall be returned to SENSIRION's factory at the Buyer's expense; and
- the warranty period for any repaired or replaced product shall be limited to the unexpired portion of the original period.

This warranty does not apply to any equipment which has not been installed and used within the specifications recommended by SENSIRION for the intended and proper use of the equipment. EXCEPT FOR THE WARRANTIES EXPRESSLY SET FORTH HEREIN, SENSIRION MAKES NO WARRANTIES, EITHER EXPRESS OR IMPLIED, WITH RESPECT TO THE PRODUCT. ANY AND ALL WARRANTIES, INCLUDING WITHOUT LIMITATION, WARRANTIES OF MERCHANTABILITY OR FITNESS FOR A PARTICULAR PURPOSE, ARE EXPRESSLY EXCLUDED AND DECLINED.

SENSIRION is only liable for defects of this product arising under the conditions of operation provided for in the data sheet and proper use of the goods. SENSIRION explicitly disclaims all warranties, express or implied, for any period during which the goods are operated or stored not in accordance with the technical specifications.

SENSIRION does not assume any liability arising out of any application or use of any product or circuit and specifically disclaims any and all liability, including without limitation consequential or incidental damages. All operating parameters, including without limitation recommended parameters, must be validated for each customer's applications by customer's technical experts. Recommended parameters can and do vary in different applications.

SENSIRION reserves the right, without further notice, (i) to change the product specifications and/or the information in this document and (ii) to improve reliability, functions and design of this product.

Headquarters and Subsidiaries

SENSIRION AG
Laubisruetistr. 50
CH-8712 Staefa ZH
Switzerland

phone: +41 44 306 40 00
fax: +41 44 306 40 30
info@sensirion.com
www.sensirion.com

Sensirion AG (Germany)
phone: +41 44 927 11 66

Sensirion Inc. USA
phone: +1 805 409 4900
info_us@sensirion.com
www.sensirion.com

Sensirion Japan Co. Ltd.
phone: +81 3 3444 4940
info-jp@sensirion.com
www.sensirion.co.jp

Sensirion Korea Co. Ltd.
phone: +82 31 337 7700~3
info-kr@sensirion.com
www.sensirion.co.kr

Sensirion China Co. Ltd.
phone: +86 755 8252 1501
info-cn@sensirion.com
http://www.sensirion.com.cn/

To find your local representative, please visit www.sensirion.com/contact

資　料　來　源　：　Sensirion　官　網　：
https://www.sensirion.com/fileadmin/user_upload/customers/sensirion/Dokumente/Humidity_Sensors/Sensirion_Humidity_Sensors_SHT3x_Datasheet_digital.pdf

HTU21D Datasheet

HTU21D(F) Sensor

Digital Relative Humidity sensor with Temperature output

√RoHS

- **DFN type package**
- **Relative Humidity and Temperature Digital Output, I²C interface**
- **Fully calibrated**
- **Lead free sensor, reflow solderable**
- **Low power consumption**
- **Fast response time**

DESCRIPTION

The HTU21D(F) is a new digital humidity sensor with temperature output by MEAS. Setting new standards in terms of size and intelligence, it is embedded in a reflow solderable Dual Flat No leads (DFN) package with a small 3 x 3 x 0.9 mm footprint. This sensor provides calibrated, linearized signals in digital, I²C format.

HTU21D(F) digital humidity sensors are dedicated humidity and temperature plug and play transducers for OEM applications where reliable and accurate measurements are needed. Direct interface with a micro-controller is made possible with the module for humidity and temperature digital outputs. These low power sensors are designed for high volume and cost sensitive applications with tight space constraints.

Every sensor is individually calibrated and tested. Lot identification is printed on the sensor and an electronic identification code is stored on the chip – which can be read out by command. Low battery can be detected and a checksum improves communication reliability. The resolution of these digital humidity sensors can be changed by command (8/12bit up to 12/14bit for RH/T).

With MEAS' improvements and miniaturization of this sensor, the performance-to-price ratio has been improved – and eventually, any device should benefit from its cutting edge energy saving operation mode.

Optional PTFE filter/membrane (F) protects HTU21D digital humidity sensors against dust and water immersion, as well as against contamination by particles. PTFE filter/membranes preserve a high response time. The white PTFE filter/membrane is directly stuck on the sensor housing.

FEATURES

- Full interchangeability with no calibration required in standard conditions
- Instantaneous desaturation after long periods in saturation phase
- Compatible with automatized assembly processes, including Pb free and reflow processes
- Individual marking for compliance to stringent traceability requirements

APPLICATIONS

- Automotive: defogging, HVAC
- Home Appliance
- Medical
- Printers
- Humidifier

Digital Relative Humidity sensor with Temperature output

NOMENCLATURE

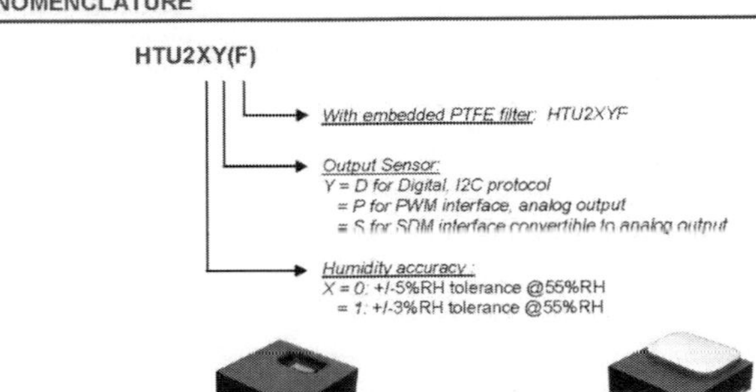

HTU2XY Modules HTU2XYF Modules

PERFORMANCE SPECS

MAXIMUM RATINGS

Ratings	Symbol	Value	Unit
Storage Temperature	T_{stg}	-40 to 125	°C
Supply Voltage (Peak)	V_{cc}	3.8V	V_{dc}
Humidity Operating Range	RH	0 to 100	%RH
Temperature Operating Range	T_a	-40 to +125	°C
VDD to GND		-0.3 to 3.6V	V
Digital I/O pins (DATA/SCK) to VDD		-0.3 to VDD+0.3	V
Input current on any pin		-10 to +10	mA

Peak conditions: less than 10% of the operating time

Exposure to absolute maximum rating conditions for extended periods may affect the sensor reliability.

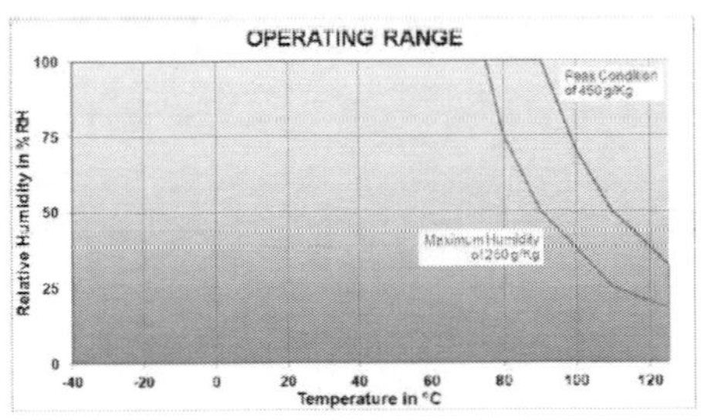

ELECTRICAL AND GENERAL ITEMS

(@T = 25°C, @Vdd = 3V)

Characteristics		Symbol	Min	Typ	Max	Unit
Voltage Supply		VDD	1.5	3.0	3.6	V
Current consumption [1]	Sleep mode	idd		0.02	0.14	μA
	Measuring		300	450	500	μA
Power Dissipation	Sleep mode			0.06	0.5	μW
	Average 8bit [2]			2.7		μW
Communication		digital 2-wire interface, I²C protocol				
Heater	VDD=3V	5.5mW/ΔT=+0.5-1.5°C				
Storage		-40°C/125°C				

[1] Conditions: V_{dd} = 3V, SCK= 400kHz at 25°C
[2] Conditions: V_{dd} = 3V, SCK= 400kHz, Temp<60°C, duty cycle <10%

SENSOR PERFORMANCE

RELATIVE HUMIDITY

(@T = 25°C, @Vdd = 3V)

Characteristics		Symbol	Min	Typ	Max	Unit
Resolution	12 bits			0.04		%RH
	8 bits			0.7		%RH
Humidity Operating Range		RH	0		100	%RH
Relative Humidity Accuracy @25°C (20%RH to 80%RH)	typ			±2		%RH
	max		See graph 1			%RH
Replacement		fully interchangeable				
Temperature coefficient (from 0°C to 80°C)		T_{cc}			-0.15	%RH/°C
Humidity Hysteresis				±1		%RH
Measuring Time [1]	12 bits			14	16	ms
	11 bits			7	8	ms
	10 bits			4	5	ms
	8 bits			2	3	ms
PSRR					±10	LSB
Recovery time after 150 hours of condensation		t		10		s
Long term drift				0.5		%RH/yr
Response Time (at 63% of signal) from 33 to 75%RH [2]		τ_{RH}		5	10	s

[1] Typical values are recommended for calculating energy consumption while maximum values shall be applied for calculating waiting times in communication.
[2] At 1m/s air flow

GRAPH 1 : RELATIVE HUMIDITY ERROR BUDGET CONDITIONS AT 25°C

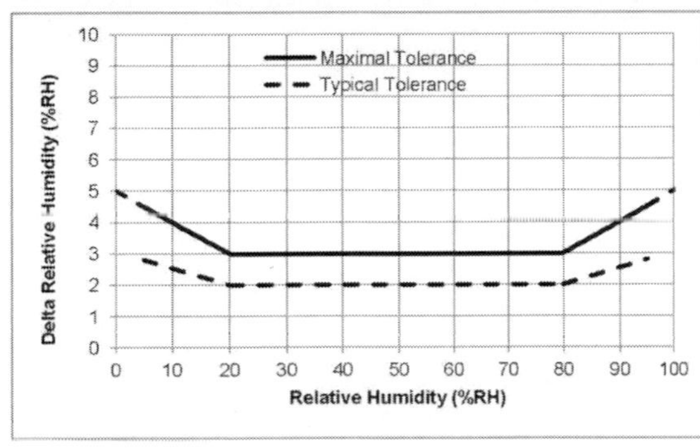

- HTU21D(F) sensors are specified for optimum accuracy measurements within 5 to 95%RH.
- Operation out of this range (< 5% or > 95% RH, including condensation) is however possible.

TEMPERATURE COEFFICIENT COMPENSATION EQUATION

Using the following temperature coefficient compensation equation will guarantee Relative Humidity accuracy given p.3, from 0°C to 80°C:

$$RH_{compensatedT} = RH_{actualT} + (25 - T_{actual}) \times CoeffTemp$$

RHactualT	Ambient humidity in %RH, computed from HTU21D(F) sensor
Tactual	Humidity cell temperature in °C, computed from HTU21D(F) sensor
CoeffTemp	Temperature coefficient of the HTU21D(F) in %RH/°C

TEMPERATURE

Characteristics		Symbol	Min	Typ	Max	Unit
Resolution	14 bit			0.01		°C
	12 bit			0.04		°C
Temperature Operating Range		T	-40		+125	°C
Temperature Accuracy @25°C	typ			±0.3		°C
	max		See graph 2			°C
Replacement			fully interchangeable			
Measuring time [1]	14 bit			44	50	ms
	13 bit			22	25	ms
	12 bit			11	13	ms
	11 bit			6	7	ms
PSSR					±25	LSB
Long term drift				0.04		°C/yr
Response Time (at 63% of signal) from 15°C to 45°C [2]		τT		10		s

[1] Typical values are recommended for calculating energy consumption while maximum values shall be applied for calculating waiting times in communication.
[2] At 1m/s air flow

GRAPH 2 : TEMPERATURE ERROR BUDGET

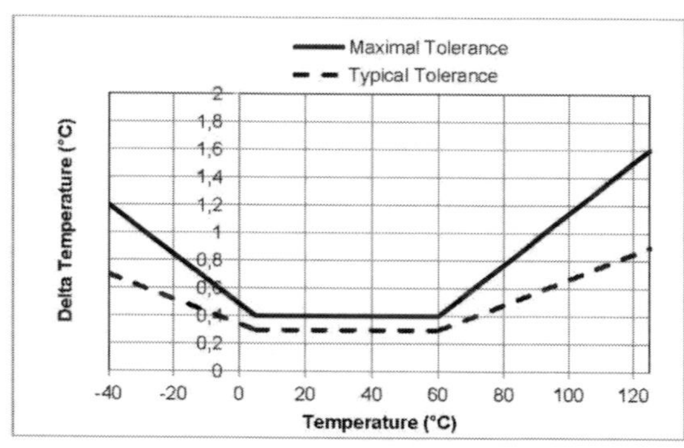

USER GUIDE HTU21D(F)

APPLICATION INFORMATION

- **Soldering instructions: Lead free reflow soldering recommended process**

For soldering HTU21D(F) sensor standard reflow soldering ovens may be used.

HTU21D(F) sensor as a humidity sensitive component (as classified by IPC/JEDEC J-STD-020 or equivalent documented procedure with peak temperature at 260°C during up to 30 seconds for Pb-free assembly in IR/convection reflow ovens) must be handled in a manner consistent with IPC/JEDEC J-STD-033 or an equivalent documented procedure. IPC-1601 provides humidity control, handling and packing of PCBs.

The HTU21D(F) sensor is qualified to withstand one lead free reflow soldering recommended process profile below according to JEDEC standard.

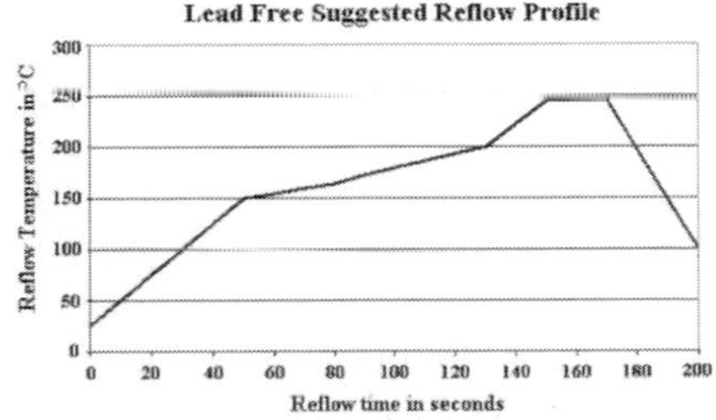

Mount parts within 24 hours after printing solder paste to avoid potential dry up.

For manual soldering, contact time must be limited to 5 seconds at up to 350°C.

For the design of the HTU21D(F) sensor footprint, it is recommended to use dimensions according to figure below.

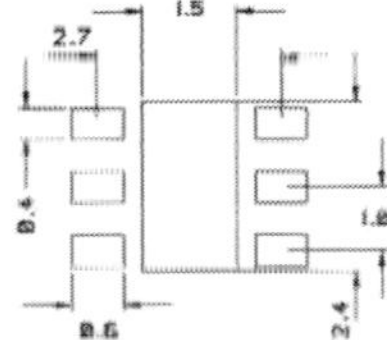

Recommended footprint for HTU21D(F) sensors. Values in mm.

No specific conditioning of devices is necessary after soldering process, either manual or reflow soldering. Optimized performance in case of metrological measurements can be reached with stabilization of devices (24 hours at 25°C / 55%RH). Similar process is advised after exposure of the devices to extreme relative humidity conditions.

In no case, neither after manual nor reflow soldering, a board wash shall be applied. Therefore, it is strongly recommended to use a "no-clean" solder paste. In case of applications with exposure of the sensor to corrosive gases or condensed water (i.e. environments with high relative humidity) the soldering pads shall be sealed (e.g. conformal coating) to prevent loose contacts or short cuts.

- **Storage Conditions and Handling Instructions**

It is recommended to store HTU21D(F) sensor in its original packaging at following conditions: Temperature shall be in the range of -40°C – 125°C.

- **Temperature Effects**

Relative humidity reading strongly depends on temperature. Therefore, it is essential to keep humidity sensors at the same temperature as the air of which the relative humidity is to be measured.

In case of testing or qualification the reference sensor and test sensor must show equal temperature to allow for comparing humidity readings.

The HTU21D(F) sensor should be mounted in a way that prevents heat transfer from electronic sensor or that keeps it as low as possible. Advice can be ventilation, reduction of copper layers between the HTU21D(F) sensor and the rest of the PCB or milling a slit into the PCB around the sensor (1mm minimum width).

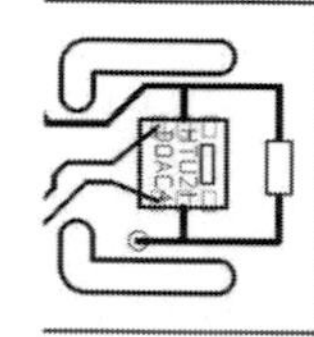

Example of HTU21D(F) sensor mounting with slits mills to minimize heat transfer

- **Materials Used for Sealing / Mounting**

For sealing and gluing (use sparingly), use high filled epoxy for electronic packaging and silicone. For any specific material please request to humidity.application@meas-spec.com.
Window must remain uncovered.

- **Wiring Considerations and Signal Integrity**

Carrying the SCK and DATA signal parallel and in close proximity (e.g. in wires) for more than 10 cm may result in cross talk and loss of communication.
This may be resolved by routing VDD and/or GND between the two data signals and/or using shielded cables. Furthermore, slowing down SCK frequency will possibly improve signal integrity.

Power supply pins (VDD, GND) must be bypassed with a 100nF capacitor if wires are used. Capacitor should be placed as close as possible to the sensor.

- **ESD (ElectroStatic Discharge)**

ESD immunity is qualified according to:

- JEDEC JESD22-A114 method (Human Body Model at ±4kV) for pads & open window
- JEDEC JESD22-A115 method (Machine Model ±200V)
- ESDA ESD-STM5.3.1-1999 and AEC-Q100-011 (charged device model, 750V corner pins, 500V other pins)

Latch-up immunity is provided at a force current of ±100mA with Tamb=25°C according to JEDEC JESD78. For exposure beyond named limits the sensor need additional protection circuit.

INTERFACE SPECIFICATION

N°	Function	Comment
1	DATA	Data bit-stream
2	GND	Ground
3	NC	Must be left unconnected
4	NC	Must be left unconnected
5	VDD	Supply Voltage
6	SCK	Selector for RH or Temp
PAD		Ground or unconnected

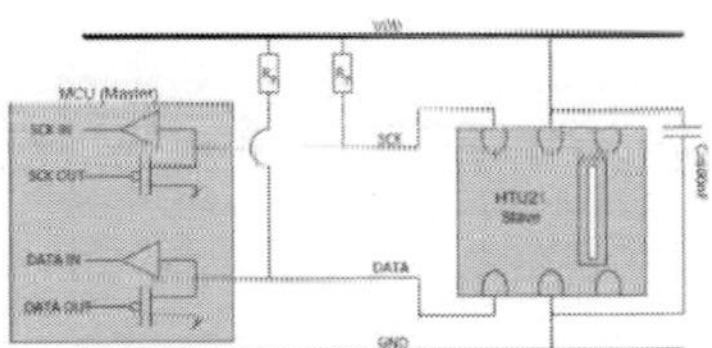

Typical application circuit, including pull-up resistor Rp and decoupling of VDD and GND by a capacitor.

- **Power Pins (VDD, GND)**

The supply voltage of HTU21D(F) sensors must be in the range of 1.5VDC - 3.6VDC. Recommended supply voltage is 3VDC (regulated).

However the typical application circuit includes a pull-up resistor R on data wire and a 100nF decoupling capacitor between VDD and GND, placed as close as possible to the sensor.

- **Serial clock input (SCK)**

SCK is used to synchronize the communication between microcontroller and HTU21D(F) sensor. Since the interface consists of fully static logic there is no minimum SCK frequency.

- **Serial data (DATA)**

The DATA pin is used to transfer data in and out of the device. For sending a command to the HTU21D(F) sensor, DATA is valid on the rising edge of SCK and must remain stable while SCK is high. After the falling edge of SCK, the DATA value may be changed. For safe communication DATA shall be valid t_{SU} and t_{HD} before the rising and after the falling edge of SCK, respectively. For reading data from the HTU21D(F) sensor, DATA is valid t_{VD} after SCK has gone low and remains valid until the next falling edge of SCK.
An external pull-up resistor (e.g. 10kΩ) on SCK is required to pull the signal high only for open collector or open drain technology microcontrollers. In most of the cases, pull-up resistors are internally included in I/O circuits of microcontrollers.

ELECTRICAL CHARACTERISTICS

- **Input/output DC characteristics**

(VDD=3V, Temperature=25°C unless otherwise noted)

Characteristics		Symbol	Min	Typ	Max	Unit
Low level output voltage	VDD=3V -4mA<IOL<0mA	VOL	0	-	0.4	V
High level output voltage		VOH	70%VDD	-	VDD	V
Low level input voltage		VIL	0	-	30%VDD	V
High level input voltage		VIH	70%VDD	-	VDD	V

- **Timing specifications of digital input/output pads for I²C fast mode**

Characteristics	Symbol	Min	Typ	Max	Unit
SCK frequency	f_{SCK}	0	-	0.4	MHz
SCK high time	t_{SCKLH}	0.6	-	-	µs
SCK low time	t_{SCLL}	1.3	-	-	µs
DATA set-up time	t_{SU}	100	-	-	ns
DATA hold-time	t_{HD}	0	-	900	ns
DATA valid-tile	t_{VD}	0	-	400	ns
SCK/DATA fall time	t_F	0	-	100	ns
SCK/DATA rise time	t_R	0	-	300	ns
Capacitive load on bus line	C_B	0	-	500	pF

- **Timing diagram for digital input/output pads**

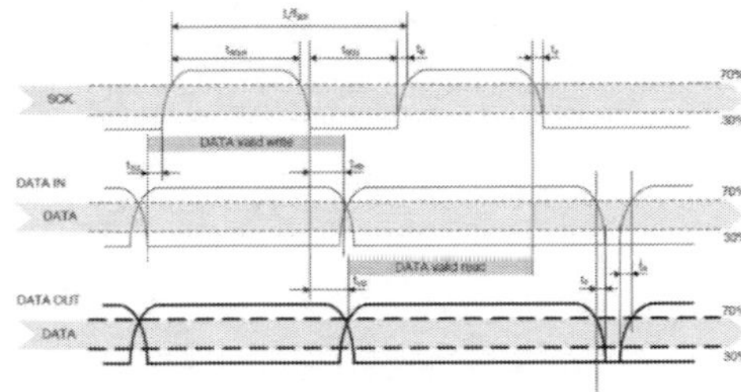

DATA directions are seen from the HTU21D(F) sensor. DATA line in bold is controlled by the sensor. DATA valid read time is triggered by falling edge of anterior toggle.

COMMUNICATION PROTOCOL WITH HTU21D(F) SENSOR

- **Start-up sensor**

The HTU21D(F) sensor requires a voltage supply between 1.5V and 3.6V. After power up, the device needs at most 15ms while SCK is high for reaching idle state (sleep mode), i.e to be ready accepting commands from the MCU. No command should be sent before that time. Soft reset is recommended at start, refer p.11.

- **Start sequence (S)**

To initiate transmission, a start bit has to be issued. It consists of a lowering of the DATA line while SCK is high followed by lowering SCK.

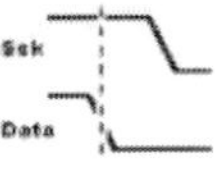

- **Stop sequence (P)**

To stop transmission, a stop bit has to be issued. It consists of a heightening of the DATA line while SCK is high preceded by a heightening of the SCK.

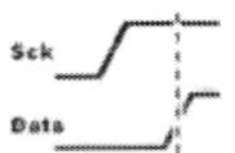

HTU21D(F) SENSOR LIST OF COMMANDS AND REGISTER ADRESSES

For sample source code, please request to humidity.application@meas-spec.com.

- **Sending a command**

After sending the start condition, the subsequent I²C header consist of a 7-bit I²C device address 0x40 and a DATA direction bit ('0' for Write access : 0x80). The HTU21D(F) sensor indicates the proper reception of a byte by pulling the DATA pin low (ACK bit) after the falling edge of the 8th SCK clock. After the issue of a measurement command (0xE3 for temperature, 0xE5 for relative humidity), the MCU must wait for the measurement to complete. The basic commands are given in the table below:

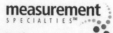

Command	Code	Comment
Trigger Temperature Measurement	0xE3	Hold master
Trigger Humidity Measurement	0xE5	Hold master
Trigger Temperature Measurement	0xF3	No Hold master
Trigger Humidity Measurement	0xF5	No Hold master
Write user register	0xE6	
Read user register	0xE7	
Soft Reset	0xFE	

- **Hold/No Hold master modes**

There are two different operation modes to communicate with the HTU21D(F) sensor: Hold Master mode and No Hold Master mode.

In the first case, the SCK line is blocked (controlled by HTU21D(F) sensor) during measurement process while in the second case the SCK line remain open for other communication while the sensor is processing the measurement.

No Hold Master mode allows for processing other I²C communication tasks on a bus while the HTU21D(F) sensor is measuring. A communication sequence of the two modes is available below.
In the Hold Master mode, the HTU21D(F) pulls down the SCK line while measuring to force the master into a wait state. By releasing the SCK line, the HTU21D(F) sensor indicates that internal processing is completed and that transmission may be continued.

In the No Hold Master mode, the MCU has to poll for the termination of the internal processing of the HTU21D(F) sensor. This is done by sending a start condition followed by the I²C header ('1' for Read access: 0x81) as shown below. If the internal processing is finished, the HTU21D(F) sensor acknowledges the poll of the MCU and data can be read by the MCU. If the measurement processing is not finished, the HTU21D(F) sensor answers no ACK bit and start condition must be issued once more.

For both modes, since the maximum resolution of the measurement is 14 bits, the two last least significant bits (LSBs, bits 43 and 44) are used for transmitting status information. Bit 1 of the two LSBs indicates the measurement type ('0': temperature, '1': humidity). Bit 0 is currently not assigned.

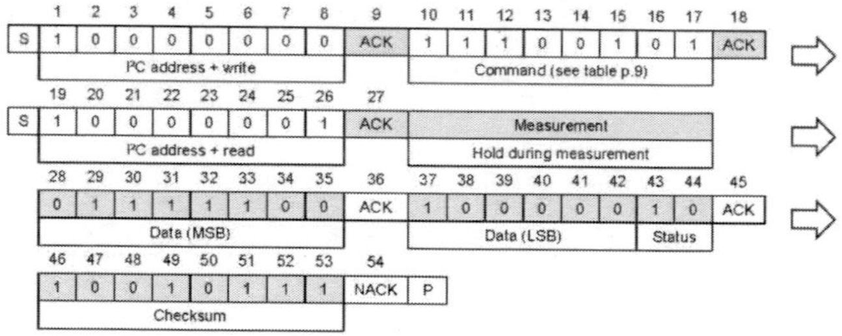

Hold Master communication sequence

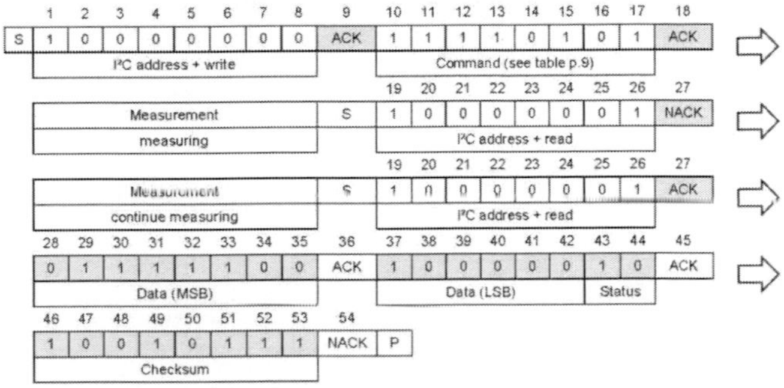

	1	2	3	4	5	6	7	8	9	10	11	12	13	14	15	16	17	18
S	1	0	0	0	0	0	0	0	ACK	1	1	1	1	0	1	0	1	ACK
	I²C address + write									Command (see table p.9)								

		19	20	21	22	23	24	25	26	27
Measurement	S	1	0	0	0	0	0	0	1	NACK
measuring		I²C address + read								

		19	20	21	22	23	24	25	26	27
Measurement	S	1	0	0	0	0	0	0	1	ACK
continue measuring		I²C address + read								

28	29	30	31	32	33	34	35	36	37	38	39	40	41	42	43	44	45
0	1	1	1	1	1	0	0	ACK	1	0	0	0	0	0	1	0	ACK
Data (MSB)									Data (LSB)						Status		

46	47	48	49	50	51	52	53	54	
1	0	0	1	0	1	1	1	NACK	P
Checksum									

No Hold Master communication sequence

Grey blocks are controlled by HTU21D(F) sensor.
For Hold Master sequence, bit 45 may be changed to NACK followed by a stop condition to omit checksum transmission.

For No Hold Master sequence, if measurement is not completed upon "read" command, sensor does not provide ACK on bit 27 (more of these iterations are possible). If bit 45 is changed to NACK followed by stop condition, checksum transmission is omitted.

In those examples, the HTU21D(F) sensor output is S_{RH} = '0111'1100'1000'0000 (0x7C80). For the calculation of physical values status bits must be set to '0'. Refer to "Conversion of signal outputs" section p.14.

The maximum duration for measurement depends on the type of measurement and resolution chosen. Maximum values shall be chosen for the communication planning of the MCU. Refer to the table p.3 and p.4 regarding measuring time specifications.

I²C communication allows for repeated start conditions without closing prior sequence with stop condition.

- **Soft reset**

This command is used for rebooting the HTU21D(F) sensor switching the power off and on again. Upon reception of this command, the HTU21D(F) sensor system reinitializes and starts operation according to the default settings with the exception of the heater bit in the user register. The soft reset takes less than 15ms.

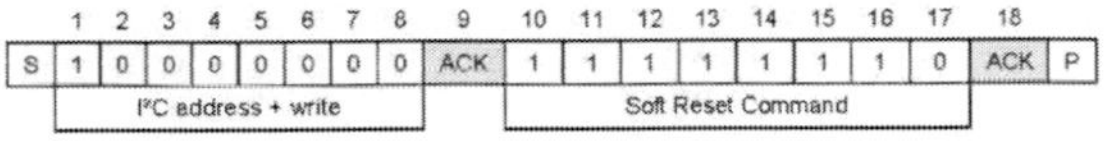

	1	2	3	4	5	6	7	8	9	10	11	12	13	14	15	16	17	18	
S	1	0	0	0	0	0	0	0	ACK	1	1	1	1	1	1	1	0	ACK	P
	I²C address + write									Soft Reset Command									

Grey blocks are controlled by HTU21D(F) sensor.

- **User register**

The content of user register is described in the table below. Reserved bits must not be changed and default values of respective reserved bits may change over time without prior notice. Therefore, for any writing to user register, default values of reserved bits must be read first.

The "End of Battery" alert/status is activated when the battery power falls below 2.25V.

The heater is intended to be used for functionality diagnosis: relative humidity drops upon rising temperature. The heater consumes about 5.5mW and provides a temperature increase of about 0.5-1.5°C.

OTP reload is a safety feature and load the entire OTP settings to the register, with the exception of the heater bit, before every measurement. This feature is disabled per default and it is not recommended for use. Please use soft reset instead as it contains OTP reload.

Bit	#Bits	Description/Coding	Default
7,0	2	Measurement resolution Bit 7 / Bit 0 / RH / Temp 0 / 0 / 12 bits / 14 bits 0 / 1 / 8 bits / 12 bits 1 / 0 / 10 bits / 13 bits 1 / 1 / 11 bits / 11 bits	'00'
6	1	Status: End of Battery[1] '0': VDD>2.25V '1': VDD<2.25V	'0'
3, 4, 5	3	Reserved	'0'
2	1	Enable on-chip heater	'0'
1	1	Disable OTP reload	'1'

[1] This status bit is updated after each measurement

Cut-off value for "End of Battery" signal may vary by ±0.1V.
Reserved bits must not be changed.
OTP reload active loads default settings after each time a measurement command is issued.

- **I²C communication reading and writing the user register example**

In this example, the resolution is set to 8 bits / 12 bits (for RH/Temp) from default configuration.

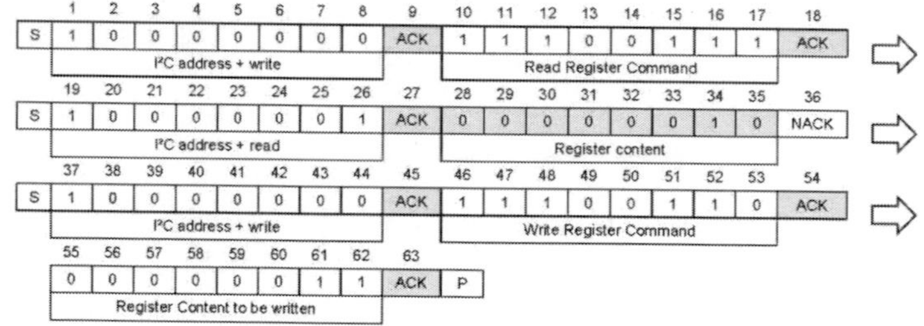

Grey blocks are controlled by HTU21D(F) sensor.

- **CRC Checksum**

HTU21D(F) sensor provides a CRC-8 checksum for error detection. The polynomial used is $X^8 + X^5 + X^4 + 1$.

Basic Considerations

CRC stands for Cyclic Redundancy Check. It is one of the most effective error detection schemes and requires a minimal amount of resources.

The types of errors that are detectable with CRC that is implemented in HTU21D(F) sensors are:

- Any odd number of errors anywhere within the data transmission
- All double-bit errors anywhere within the transmission
- Any cluster of errors that can be contained within an 8-bit window (1-8 bits incorrect)
- Most larger clusters of errors

A CRC is an error-detecting code commonly used in digital networks and storage devices to detect accidental changes to raw data.

Blocks of data entering these systems get a short check value attached, based on the remainder of a polynomial division of their contents; on retrieval the calculation is repeated, and corrective action can be taken against presumed data corruption if the check values do not match.

CRCs are so called because the check (data verification) value is a redundancy (it expands the message without adding information) and the algorithm is based on cyclic codes. CRCs are popular because they are simple to implement in binary hardware, easy to analyze mathematically, and particularly good at detecting common errors caused by noise in transmission channels. Because the check value has a fixed length, the function that generates it is occasionally used as a hash function.

CRC for HTU21D(F) sensors using I²C Protocol

When HTU21D(F) sensors are run by communicating with the standard I²C protocol, an 8-bit CRC can be used to detect transmission errors. The CRC covers all read data transmitted by the sensor. CRC properties for HTU21D(F) sensors communicating with I²C protocol are listed in the table below.

CRC with I²C protocol	
Generator polynomial	$X^8 + X^5 + X^4 + 1$
Initialization	0x00
Protected data	Read data
Final Operation	none

CRC calculation

To compute an n-bit binary CRC, line the bits representing the input in a row, and position the (n+1)-bit pattern representing the CRC's divisor (called a "polynomial") underneath the left-hand end of the row.

This is first padded with zeroes corresponding to the bit length n of the CRC.

If the input bit above the leftmost divisor bit is 0, do nothing. If the input bit above the leftmost divisor bit is 1, the divisor is XORed into the input (in other words, the input bit above each 1-bit in the divisor is toggled). The divisor is then shifted one bit to the right, and the process is repeated until the divisor reaches the right-hand end of the input row.

Since the left most divisor bit zeroed every input bit it touched, when this process ends the only bits in the input row that can be nonzero are the n bits at the right-hand end of the row. These n bits are the remainder of the division step, and will also be the value of the CRC function.

The validity of a received message can easily be verified by performing the above calculation again, this time with the check value added instead of zeroes. The remainder should equal zero if there are no detectable errors.

CRC examples

The input message 11011100 (0xDC) will have as result 01111001 (0x79).

The input message 01101000 00111010 (0x683A: 24.7°C) will have as result 01111100 (0x7C).

The input message 01001110 10000101 (0x4E85: 32.3%RH) will have as result 01101011 (0x6B).

CONVERSION OF SIGNAL OUTPUTS

Default resolution is set to 12-bit relative humidity and 14-bit temperature readings. Measured data are transferred in two byte packages, i.e. in frames of 8-bit length where the most significant bit (MSB) is transferred first (left aligned). Each byte is followed by an acknowledge bit. The two status bits, the last bits of LSB, must be set to '0' before calculating physical values.

To accommodate/adapt any process variation (nominal capacitance value of the humidity die), tolerances of the sensor above 100%RH and below 0%RH must be considered. As a consequence:

- 118%RH corresponds to 0xFF which is the maximum RH digital output that can be sent out from the ASIC. RH output can reach 118%RH and above this value, there will have a clamp of the RH output to this value.
- -6%RH corresponds to 0x00 which is the minimum RH digital output that can be sent out from the ASIC. RH output can reach -6%RH and below this value, there will have a clamp of the RH output to this value.

- **Relative Humidity conversion**

With the relative humidity signal output S_{RH}, the relative humidity is obtained by the following formula (result in %RH), no matter which resolution is chosen:

$$RH = -6 + 125 \times \frac{S_{RH}}{2^{16}}$$

In the example given p.10, the transferred 16-bit relative humidity data is 0x7C80: 31872. The relative humidity results to be 54.8%RH.

- **Temperature conversion**

The temperature T is calculated by inserting temperature signal output S_{Temp} into the following formula (result in °C), no matter which resolution is chosen:

$$Temp = -46.85 + 175.72 \times \frac{S_{Temp}}{2^{16}}$$

APPLICATION: DEW POINT TEMPERATURE MEASUREMENT

The dew point is the temperature at which the water vapor in the air becomes saturated and condensation begins.

The dew point is associated with relative humidity. A high relative humidity indicates that the dew point is closer to the current air temperature. Relative humidity of 100% indicates that the dew point is equal to the current temperature (and the air is maximally saturated with water). When the dew point stays constant and temperature increases, relative humidity will decrease.

Dew point temperature of the air is calculated using Ambient Relative Humidity and Temperature measurements from HTU21D(F) sensor with following formulas given below:

Partial Pressure (PP_{Tamb}) formula from Ambient Temperature:

$$PP_{Tamb} = 10^{\left[A - \frac{B}{(Tamb + C)}\right]}$$

Dew point Temperature (T_d) formula from Partial Pressure (PP_{Tamb}):

$$T_d = -\left[\frac{B}{\log_{10}\left(RH_{amb} \times \frac{PP_{Tamb}}{100}\right) - A} + C\right]$$

PP_{Tamb}	Partial Pressure in mmHg at ambient temperature (T_{amb})
RH_{amb}	Ambient humidity in %RH, computed from HTU21D(F) sensor
T_{amb}	Humidity cell temperature in °C, computed from HTU21D(F) sensor
T_d	Calculated Dew Point in °C
A, B, C	Constants: A=8.1332; B=1762.39; C=235.66

PACKAGE OUTLINE

- **HTU21D Sensor Dimensions**

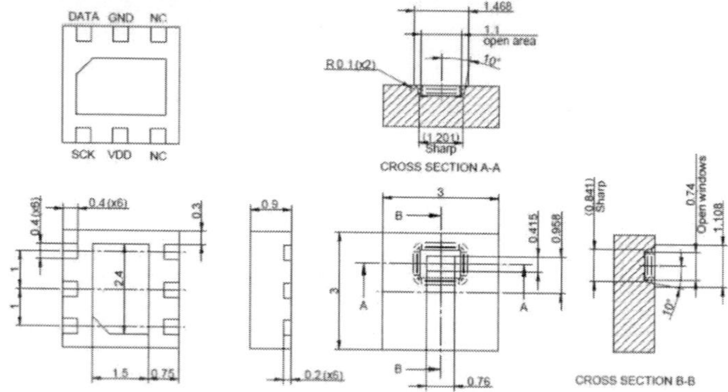

- **HTU21DF Sensor Dimensions**

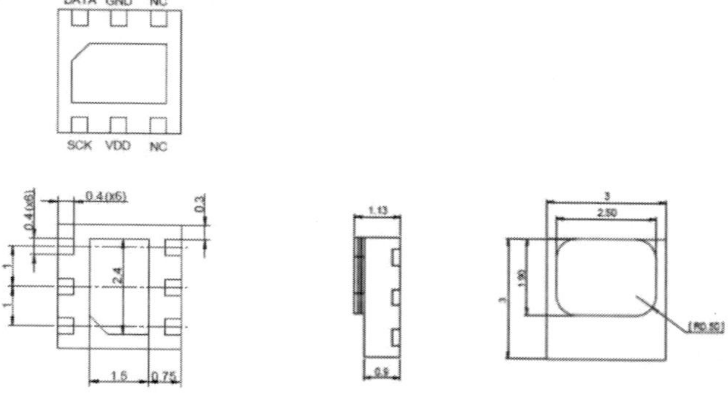

Dimensions are given in mm, tolerances are ±0.1mm. The die pad (thermal center pad) is internally connected to GND.

- **Packaging Type**

HTU21D(F) sensors are provided in DFN packaging. DFN stands for Dual Flat No leads.

The HTU21D(F) sensor chip is mounted to a lead frame made of Cu and plated with Ni/Pd/Au. Chip and lead frame are over molded by green epoxy-based mold compound. Please note that side walls of sensors are diced and hence lead frame at diced edge is not covered with respective protective coating.

The total weight of the sensor is 0.025g.

- **Traceability Information**

All HTU21D(F) sensors are laser marked with an alphanumeric, five-digit code on the sensor as pictured below.

The marking on the HTU21D(F) sensor consists of two lines with five digits each:

- The first line denotes the sensor type: HTU21.
- The second line denotes several information as:
    - The first digit of the second line defines the output mode:
        - D = digital and I²C
        - P = PWM
        - S = SDM
    - The second digit defines the manufacturing year: 2 = 2012, 3 = 2013, etc.
    - The last three digits represent an alphanumeric tracking code. That code can be decoded by MEAS only and allows for tracking on batch level through production, calibration and testing and will be provided upon justified request.

Laser marking on HTU21D(F) sensor

Reels are also labeled, as displayed below and give additional traceability information.

With:

XX:	Sensor Type (21 for HTU21D(F))
O:	Output mode (D = Digital, P = PWM, S = SDM)
(F):	Sensor with PTFE membrane (only for HTU21DF)
NN:	Product revision number
TTTTTTTTT:	MEAS Traceability Code
YY:	Two last digits of the year
DDD:	Day of the year
QQQQ:	Quantity per real (400, 1500 or 5000 units)

- **Tape and Reel Packaging**

HTU21D(F) sensors are shipped in tape & reel packaging, sealed into antistatic ESD bags.

Standard packaging sizes are 400, 1500 and 5000 units per reel. Each reel contains 440mm (55 pockets) header tape and 200mm (25 pockets) trailer tape. The drawing of the packaging tapes with sensor orientation is shown in the picture below.

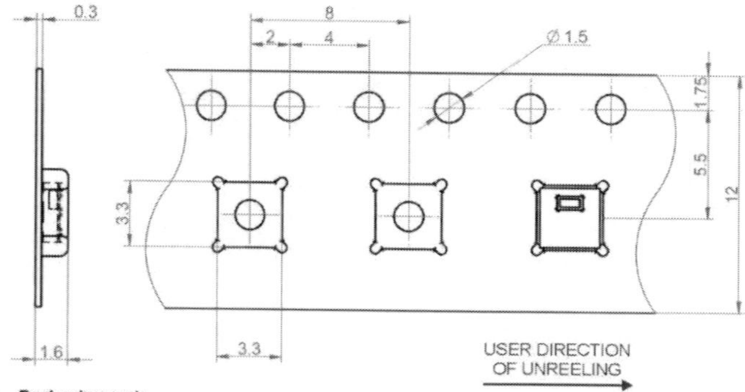

- **Packaging reels**

For 400 and 1500 units: outside diameter of 7" (178mm) and a 1/2" (13mm) diameter arbor hole.

For 5000 units: outside diameter of 13" (330mm) and a 1/2" (13mm) diameter arbor hole.

ORDERING INFORMATION

**** HTU21D – DIGITAL TEMPERATURE AND RELATIVE HUMIDITY MODULE ****

PACKAGE: TAPE AND REEL M.P.Q OF 400 PIECES, 1500 PIECES OR 5000 PIECES

- **HPP828E031R4 - HTU21D in tape and reel of 400 pieces**
- **HPP828E031R1 - HTU21D in tape and reel of 1500 pieces**
- **HPP828E031R5 - HTU21D in tape and reel of 5000 pieces**

**** HTU21DF – DIGITAL TEMPERATURE AND RELATIVE HUMIDITY MODULE WITH PTFE MEMBRANE ****

PACKAGE: TAPE AND REEL M.P.Q OF 400 PIECES, 1500 PIECES OR 5000 PIECES

- **HPP828E131R4 - HTU21DF in tape and reel of 400 pieces**
- **HPP828E131R1 - HTU21DF in tape and reel of 1500 pieces**
- **HPP828E131R5 - HTU21DF in tape and reel of 5000 pieces**

**** HTU21D DEMOKIT – HPP828KIT ****

This is a USB device for MEAS Model HTU21D Digital Relative Humidity & Temperature sensor demonstration. Supporting up to 4 sensor acquisitions at the same time, it shows the consistency of different sensors and test sensor functions conveniently.
For detailed information, please request to humidity.application@meas-spec.com.

Customer Service contact details

Measurement Specialties, Inc - MEAS France
Impasse Jeanne Benozzi
CS 83 163
31027 Toulouse Cedex 3
FRANCE
Tel:+33 (0)5 820.822.02
Fax:+33 (0)5.820.821.51
Sales: humidity.sales@meas-spec.com

MEAS Website: http://www.meas-spec.com/humidity-sensors.aspx

Revision	Comments	Who	Date
0	Document creation	D. LE GALL	April 12
A	General update	D. LE GALL-ZIRILLI	February 13
2	HTU21DF product with embedded PTFE membrane reference added, Storage conditions after soldering process updated (typing error), ESD performances updated, complementary information on RH output signal in "Conversion of signal outputs" paragraph, information on tape and reel packaging added, HTU21D demokit availability information added.	D. LE GALL-ZIRILLI	July 13
3	Correction of I²C communication reading and writing, correction of soldering peak temperature	M.ROBERT	October 2013

資料來源：Adafruit 官網：

https://cdn-shop.adafruit.com/datasheets/1899_HTU21D.pdf

參考文獻

曹永忠, 许智诚, & 蔡英德. (2014). *Arduino 电风扇设计与制作: Using Arduino to Develop a Controller of the Electric Fan*. 台湾、彰化: 渥玛数位有限公司.

曹永忠, 許智誠, & 蔡英德. (2013). *Arduino 電風扇設計與製作: The Design and Development of an Electronic Fan by Arduino Technology* (初版 ed.). 台灣、彰化: 渥瑪數位有限公司.

曹永忠, 許智誠, & 蔡英德. (2014a). *Arduino 饮水机电子控制器开发: Using Arduino to Develop a Controller of Drinking Fountain*. 台湾、彰化: 渥瑪數位有限公司.

曹永忠, 許智誠, & 蔡英德. (2014b). *Arduino 飲水機電子控制器開發: The Development of a Controller for Drinking Fountain Using Arduino*. 台灣、彰化: 渥瑪數位有限公司.

曹永忠, 許智誠, & 蔡英德. (2015a). *Arduino 程式教學(入門篇):Arduino Programming (Basic Skills & Tricks)* (初版 ed.). 台湾、彰化: 渥玛数位有限公司.

曹永忠, 許智誠, & 蔡英德. (2015b). *Arduino 程式教學(常用模組篇):Arduino Programming (37 Sensor Modules)* (初版 ed.). 台湾、彰化: 渥玛数位有限公司.

曹永忠, 許智誠, & 蔡英德. (2015c). *Arduino 编程教学(常用模块篇):Arduino Programming (37 Sensor Modules)* (初版 ed.). 台湾、彰化: 渥玛数位有限公司.

曹永忠, 許智誠, & 蔡英德. (2015d). *Arduino 编程教学(入门篇):Arduino Programming (Basic Skills & Tricks)* (初版 ed.). 台湾、彰化: 渥玛数位有限公司.

曹永忠, 許智誠, & 蔡英德. (2015e). Maker 物聯網實作：用 DHx 溫濕度感測模組回傳天氣溫溼度. *物聯網*. Retrieved from http://www.techbang.com/posts/26208-the-internet-of-things-daily-life-how-to-know-the-temperature-and-humidity

曹永忠, 許智誠, & 蔡英德. (2016a). *Arduino 空气盒子随身装置设计与开发(随身装置篇):Using Arduino to Develop a Timing Controlling Device via Internet* (初版 ed.). 台湾、彰化: 渥瑪數位有限公司.

曹永忠, 許智誠, & 蔡英德. (2016b). *Arduino 空氣盒子隨身裝置設計與開發(隨身裝置篇):Using Arduino Nano to Develop a Portable PM 2.5 Monitoring Device* (初版 ed.). 台湾、彰化: 渥瑪數位有限公司.

曹永忠, 許智誠, & 蔡英德. (2016c). *Arduino 程式教學(基本語法*

篇):Arduino Programming (Language & Syntax) (初版 ed.). 台湾、彰化: 渥瑪數位有限公司.

曹永忠, 許智誠, & 蔡英德. (2016d). *Arduino 程序教学(基本语法篇) :Arduino Programming (Language & Syntax)* (初版 ed.). 台湾、彰化: 渥瑪數位有限公司.

曹永忠, 許碩芳, 許智誠, & 蔡英德. (2015a). *Arduino 程式教學(RFID 模組篇):Arduino Programming (RFID Sensors Kit)* (初版 ed.). 台湾、彰化: 渥瑪數位有限公司.

曹永忠, 許碩芳, 許智誠, & 蔡英德. (2015b). *Arduino 編程教学(RFID 模块篇):Arduino Programming (RFID Sensors Kit)* (初版 ed.). 台湾、彰化: 渥瑪數位有限公司.

Arduino 程式教學（溫溼度模組篇）

Arduino Programming (Temperature& Humidity Modules)

作　　者：曹永忠、許智誠、蔡英德
發 行 人：黃振庭
出 版 者：崧燁文化事業有限公司
發 行 者：崧燁文化事業有限公司
E-mail：sonbookservice@gmail.com
粉 絲 頁：https://www.facebook.com/sonbookss/
網　　址：https://sonbook.net/
地　　址：台北市中正區重慶南路一段六十一號八樓 815 室
Rm. 815, 8F., No.61, Sec. 1, Chongqing S. Rd., Zhongzheng Dist., Taipei City 100, Taiwan
電　　話：(02) 2370-3310
傳　　真：(02) 2388-1990
印　　刷：京峯彩色印刷有限公司（京峰數位）
律師顧問：廣華律師事務所 張珮琦律師

定　　價：300 元
發行日期：2022 年 03 月第一版
◎本書以 POD 印製

國家圖書館出版品預行編目資料

Arduino 程式教學. 溫溼度模組篇 = Arduino programming(temperature & humidity modules) / 曹永忠, 許智誠, 蔡英德著. -- 第一版. -- 臺北市 : 崧燁文化事業有限公司, 2022.03
面；　公分
POD 版
ISBN 978-626-332-078-9(平裝)
1.CST: 微電腦 2.CST: 電腦程式語言
471.516　111001396

官網

臉書